Compendium of Rose Diseases

Prepared by
R. Kenneth Horst

American Phytopathological Society

In cooperation with

Department of Plant Pathology
Cornell University

What's in a name? That which we call a rose
By any other name would smell as sweet

> Shakespeare,
> *Romeo and Juliet*

Financial Sponsor
The Fred C. Gloeckner Foundation, Inc.

Front cover: Photograph by Howard H. Lyon, Department of Plant Pathology, Cornell University. Roses arranged by Raymond T. Fox, Department of Floriculture and Ornamental Horticulture, Cornell University
Back cover: Rose production fields of Jackson & Perkins Co., Wasco, CA. Photograph by R. Kenneth Horst with help from Howard H. Lyon

Library of Congress Catalog Card Number: 83-71230
International Standard Book Number: 0-89054-052-7

©1983 by The American Phytopathological Society
Second Printing, 1986

All rights reserved.
No part of this book may be reproduced in any form
by photocopy, microfilm, retrieval system, or any other means,
without written permission from the publisher.

Printed in the United States of America

The American Phytopathological Society
3340 Pilot Knob Road
St. Paul, Minnesota 55121, USA

To Dr. Cynthia Westcott (1898–1983),

"The Plant Doctor," a pioneer leader in plant pathology, and author of *Anyone Can Grow Roses*. Her greatest enthusiasm was always the rose.

Preface

This Compendium is the first in The American Phytopathological Society's series of disease compendia to cover diseases of a commercial flowering ornamental. Its primary purpose is to aid plant pathologists and other agricultural workers in the diagnosis of rose diseases. It is also designed to serve as a practical reference and resource for rose growers, students, researchers, educators, crop disease consultants, advisors in state and federal government and regulatory agencies, agribusiness representatives, the pest control industry, county agricultural and cooperative extension agents, and area crop specialists throughout the world.

The Compendium is amply illustrated and uses descriptive terminology. Diseases are arranged according to causal agents; infectious diseases, caused by fungi, bacteria, viruses, and nematodes, are covered first, followed by noninfectious diseases, including physiologic problems, environmental imbalances, air pollution, pesticide toxicity, and nutritional deficiencies and toxicities. A glossary, index, and selected references are included. The references cited in each section were selected to include the most contemporary literature applicable to the discussion of the disease and thus are not all-inclusive.

Discussions of disease control measures purposely stress principles and cultural practices that are not likely to become obsolete. For specific information concerning chemical control measures, the reader should consult recent literature and extension or advisory plant pathologists.

Comments regarding the general usefulness of or omissions in this Compendium are most welcome. With your suggestions, future editions can be made even more useful and practical.

The author wishes to thank the members of the Advisory and Reviewing Committee, who reviewed all or part of the manuscript, made numerous valuable suggestions, and contributed illustrations:

D. L. Coyier, Oregon State University, Corvallis
P. Fry, Massey University, New Zealand
F. A. Hakkaart, Research Station for Floriculture, Aalsmeer, The Netherlands
E. L. Halk, USDA-ARS, Beltsville, MD
M. B. Harrison, Cornell University, Ithaca, NY
C. Harwood, Jackson & Perkins Co., Wasco, CA
R. W. Langhans, Cornell University, Ithaca, NY
K. Milne, Massey University, New Zealand
L. W. Moore, Oregon State University, Corvallis
L. P. Nichols, The Pennsylvania State University, University Park
G. Nyland, University of California, Davis
K. Ohkawa, Kanagawa-Ken, Japan
N. Paludan, The State Plant Pathology Institute, Lyngby, Denmark
U. Paz, Ministry of Agriculture, Hedera Region, Israel
C. C. Powell, The Ohio State University, Columbus
M. N. Rogers, University of Missouri, Columbia
B. Scarborough, Conard-Pyle Co., West Grove, PA
G. A. Secor, North Dakota State University, Fargo
J. G. Seeley, Cornell University, Ithaca, NY
L. L. Stubbs, University of Melbourne, Australia
B. J. Thomas, Glasshouse Crops Research Institute, Littlehampton, England
V. A. Wager, South African Garden & Home, Durban, South Africa

The Department of Plant Pathology at Cornell University provided office, library, and communication facilities and secretarial and photographic services and support. The stenographic expertise of Roxanna E. Barnum and her patience and attention to detail in manuscript preparation were invaluable. The professional photographic expertise of H. H. Lyon is gratefully acknowledged. The added research assistance of Stanley Kawamoto and Karen Weaber while the manuscript was being prepared is greatly appreciated. They, along with Sek Man Wong and Robert McGovern, also helped proofread the manuscript and made critical suggestions. Carl Whittaker prepared illustrations of life cycles and drawings of fungal fruiting structures, and R. P. Korf provided valuable advice in the mycological area. H. H. Lyon and R. T. Fox helped prepare the book's cover.

The help of Marlin N. Rogers, University of Missouri, who wrote the section on noninfectious diseases, is greatly appreciated. Without the assistance of The Fred C. Gloeckner Foundation, the project could not have been completed. The invaluable aid and support of J. Shibata, Cherry City Nursery and California-Florida Plant Corporation, in preparing illustrations used in the Compendium are gratefully acknowledged. Unless otherwise indicated in the caption, illustrations are courtesy of the Department of Plant Pathology, Cornell University.

Contents

Introduction
- **2 Rose Diseases**
- 2 Infectious Diseases
- 2 Noninfectious Diseases
- 3 Disease Control

Part I. Infectious Diseases
- **5 Diseases Caused by Fungi**
- 5 Powdery Mildew
- 7 Black Spot
- 11 Rust
- 12 Verticillium Wilt
- 13 Downy Mildew
- 15 Brand Canker
- 15 Common Canker (Rose Graft Canker)
- 16 Brown Canker
- 17 Black Mold
- 18 Botrytis Blight
- 19 Canker (Dieback)
- 20 Miscellaneous Diseases Caused by Fungi
- **23 Diseases Caused by Bacteria**
- 23 Crown Gall
- 25 Hairy Root
- **26 Diseases Caused by Viruses**
- 26 Rose Mosaic
- 27 Strawberry Latent Ringspot Virus
- 28 Rose Streak
- 28 Rose Rosette
- 28 Rose Ring Pattern
- 29 Rose Wilt
- 29 Rose Spring Dwarf
- 29 Rose Leaf Curl
- 30 Rose Flower Break
- 30 Rose Flower Proliferation
- 30 Tobacco Streak Virus
- **30 Diseases Caused by Nematodes**

Part II. Noninfectious Diseases
- **33 Physiologic Problems**
- 33 Blindness
- 33 Bullheads
- 34 Bent Neck
- **34 Environmental Imbalances**
- 34 Petal Blackening in Baccara Roses
- 34 Petal Bluing in Baccara Roses
- 35 Heat and Moisture Stress
- 35 Oxygen Deficiency
- **35 Air Pollution**
- 35 Fluoride
- 35 Ethylene
- 36 Mercury Vapor
- 36 Paint Volatiles
- **36 Pesticide Toxicity**
- **36 Nutritional Deficiencies**
- **38 Nutritional Toxicities**
- 39 Salinity

41 Glossary

47 Index

Color Plates (following page **22**)

Introduction

The rose is the most popular garden plant in the world as well as the most important commercial cut flower grown under glass. Commercial cut flower roses in the United States have exceeded all other cut flowers in wholesale value in the last decade. In 1970, the wholesale value of roses was $54 million; by 1980, the wholesale value had nearly doubled to $105.7 million. In addition, roses with a wholesale value of $11 million were imported into the United States in 1980.

The modern cultivated rose in the genus *Rosa* is a monument to the achievements of practical plant breeders. The genus has differentiated into at least 200 botanical species widely distributed in the Northern Hemisphere. All members of the genus are important in horticulture for blooms as well as ornamental shrubbery.

Fossil specimens indicate the presence of primitive roses in Colorado and Oregon more than 30 million years ago. There is evidence that roses were first cultivated 4,000 to 5,000 years ago in northern Africa. The active speciation that has occurred since these early roses has resulted in many groups and hybrids that are difficult to separate taxonomically. Some well-known groups commonly referred to are briefly described in the following paragraphs.

Hybrid perpetual roses are of mixed ancestry; *R. borboniana* is the prevailing species. These cultivars have upright growth and produce large, fragrant double flowers in early summer and sparingly in fall. Most hybrid perpetual cultivars are hardy in the northern United States.

Hybrid tea roses are mostly *R. dilecta* and represent crosses between hybrid perpetuals and virtually all other rose groups. They are not as hardy as hybrid perpetuals. Hybrid teas have a recurrent and fragrant blooming habit and are a predominant garden and glasshouse rose today.

Polyantha roses, an important branch of the multiflora class, are a cross of *R. multiflora* and *R. chinensis*, correctly known botanically as *R.* × *rehderiana*. This group includes dwarf roses, which are sometimes called "baby ramblers," and a subgroup that because of its blooming habit has been commercially called the floribunda.

Tea roses are principally *R. odorata* and are used much less as a group than the hybrid teas. This group is noted for its abundance of large flowers but is not dependably hardy north of the Mason-Dixon line in the United States.

China and *Bengal* roses are essentially *R. chinensis* and *R. odorata*, and their hybrids are used in much the same way as tea roses. Some are hardy in the central United States. The blooms are generally small and red.

Noisette or *Champney* roses are hybrids of *R. noisettiana*. These excellent bush roses are not dependably hardy north of the Mason-Dixon line. This group is grown extensively in Europe.

Multiflora roses are *R. multiflora* and include climbing roses, which are often called "ramblers." The true multiflora roses have cluster flowers.

Wichuraiana roses are *R. wichuraiana* and are also called the Memorial rose. These roses are more or less evergreen and are ground trailers. They are a dominant form of hardy climbing rose and are difficult to distinguish from multiflora roses.

Manetti is not considered a group of roses but must be mentioned since it is often used as an understock for grafted roses. Its origin is somewhat confused, but older historical literature indicates that it should be called *R. manetti*. *R. manetti* is of hybrid origin and is possibly segregated from a population of *R. chinensis*. Manetti is generally known to be tolerant of shallow or dry soils and confined root systems.

Interspecific hybridization is very important in the origin of modern rose types. Most cultivated roses are derived from hybrids involving several different wild species of roses. However, serious technical difficulties caused by sterilities from species of $2n$, $3n$, $4n$, $5n$, $6n$, and $8n$ chromosomal constituencies have prevented the full utilization of the 200 or more species of *Rosa*. Advances in molecular genetics should greatly aid in overcoming this problem and should allow rapid development of improved cultivars of roses in the future.

The system of cultivar classification has become complicated and inexact as hybridization has proceeded. Selection and hybridization have given rise to some 20,000 cultivars. More than 12,000 of these are recorded in *Modern Roses 7*, published by The McFarland Company under the auspices of the American Rose Society, which serves as the international registration authority for roses. It is now nearly impossible to accurately classify new introductions on sight, because the class specifications are imprecise and often overlap. Because of this inconsistency, the World Federation of Rose Societies is attempting to develop a simplified and internationally accepted classification system. The proposed system is based on habit of growth (climbing [I] vs. nonclimbing [II]) and flowering habit (recurrent [A] vs. nonrecurrent [B]). The five classes of recurrent-flowering roses (II-A) are shrub roses, large-flowered roses (hybrid teas and most grandifloras), cluster-flowered roses (floribunda), polyanthas, and miniature roses. This classification system was approved at the first World Rose Convention in 1971 but will take time to become refined and internationally used.

Pruning is necessary to obtain good roses. In cold climates, winter injury necessitates pruning to remove dead wood and small, weak cane growth in the spring. In milder climates, dead wood and weak wood are also removed, depending on the size of plants desired. In general, pruning should be done in late winter or early spring, although some pruning may be necessary during the summer to maintain tidy plant appearance.

Roses may be propagated from rooted cuttings, by budding, or by grafting. For the gardener, propagation by rooting cuttings is the most practical technique. The simplest method is to make cuttings 12–20 cm long from stems that have just finished flowering. These cuttings are placed in the ground under a jar in a shady location and kept moist at all times. Commercial nurseries propagate roses by budding onto an understock in the field. Many cultivars are more vigorous when budded on an understock than when grown on their own roots.

Rose Diseases

Infectious Diseases

Infectious diseases are caused by fungi, bacteria, viruses, and nematodes. Disease development depends on the presence of a virulent pathogen, a susceptible host, and the proper environment. Disease symptoms may be classified as necrosis, or death of cells, tissues, or organs; hypoplasia, resulting in dwarfing or stunting; or hyperplasia, resulting in an overgrowth of plant tissue, as found in crown gall.

Fungi

Fungi are simple, filamentous plants that contain no chlorophyll and reproduce by sexual or asexual spores. More than 100,000 species of fungi have been described, of which about 20,000 species are pathogenic to plants, animals, or both. Fungi are identified by spore morphology and the mechanism of spore production. Fungus spores are easily disseminated by air currents, splashing water, and the activities of people; they can also move on or in plant parts or animals. Fungi overwinter on and in living or dead plants, in soil, and occasionally in insects.

Bacteria

Bacteria include some 4,000 procaryotic species, most of which contain no chlorophyll. Several hundred species are pathogenic to plants, animals, or both. Plant-pathogenic bacteria are generally unicellular, non-spore-forming rods. The identification of bacteria can be very difficult; it involves special staining techniques, biochemical tests, and serology. Five genera of true bacteria cause plant disease: *Pseudomonas, Xanthomonas, Agrobacterium, Erwinia,* and *Corynebacterium.* Only two of these genera—*Agrobacterium* and *Corynebacterium*—contain species known to be pathogenic to roses.

Bacteria enter plants through wounds, stomata, water pores, flower nectaries, and possibly lenticels. They can survive inactively for months in plant tissues and sometimes for years in soil. Bacteria are disseminated by people, insects and nematodes, splashing water, and windblown soil or sand. Free moisture and moderate to warm temperatures are generally required for disease development.

Viruses

Six hundred or more plant diseases are thought to be caused by viruses; however, two-thirds of these have not been conclusively proved to have a viral causal agent. Viroids, mycoplasmas, and spiroplasmas have now been shown to cause some of these diseases, but the symptomatology associated with the diseases caused by these agents is similar to that associated with diseases known to be caused by viruses. Symptoms of virus diseases include suppression of chlorophyll production, leaf mottling, yellow or necrotic ring patterns in leaves, stunting, leaf and flower distortion, witches'-broom or rosettes, and sometimes necrosis.

Viruses contain two major components: nucleic acid surrounded by a protein coat. Viroids consist solely of a small amount of RNA with no protein coat. Mycoplasmas that cause plant disease are commonly placed in two groups, namely mycoplasmalike organisms (MLOs) and spiroplasmas. MLOs are pleomorphic agents bounded by single membranes and devoid of cell walls. Spiroplasmas possess all the characteristics of MLOs but in addition have a distinct helical morphology and rotary and flexuous motility. Thus far no viroids, MLOs, or spiroplasmas have been reported to be pathogens of rose, although some of the diseases classified as virus diseases in this Compendium have not yet been conclusively determined to have a viral causal agent.

Viruses are differentiated and identified by host specificity, physical properties, purification, electron microscopy, electrophoresis, serology, and possibly other techniques. Viruses are transmitted by insects, mites, fungi, nematodes, mechanical means, grafting, dodder, and occasionally seed.

Nematodes

Nematodes are unsegmented roundworms that inhabit fresh and salt water, decaying organic matter, soil, plants, and animals throughout the world. More than 15,000 species of nematodes have been described, and about 2,200 species are parasitic on plants. Plant-parasitic nematodes are typically microscopic, transparent, mobile, and vermiform. Nematodes move passively in water, soil, and infected plant parts and on tools, vehicles, and machinery carrying infested soil; they are also disseminated by wind and many animals.

Noninfectious Diseases

Noninfectious diseases are caused by an excess, deficiency, or imbalance of nutrients; by water, pH, or environmental extremes; and by air pollutants, pesticides, and other injuries. Symptoms of noninfectious diseases may often be confused with those caused by fungi, bacteria, viruses, and nematodes.

Disease Control

Control measures of a general nature are described for each disease covered in this Compendium. Perfect disease control is rare, but economic control, such that the increased yield more than covers the cost of labor and materials, is quite possible. Disease control may be achieved by a single procedure but more often involves the integrated use of several measures. The five fundamental principles of control are exclusion, eradication, protection, resistance, and therapy.

Exclusion means preventing pathogens from entering and becoming established in uninfested fields, gardens, states, or countries. To this end, commercial growers and gardeners should use plants certified to be free of selected diseases. States and countries often establish quarantines to exclude dangerous pathogens from entry.

Eradication means elimination of the pathogen once it has become established on plants in the field or garden. Eradication can be accomplished by removing diseased specimens, by rotating susceptible crops with nonsusceptible crops, and by disinfecting, usually with chemicals.

Protection is the interposition of some protective barrier between the susceptible part of the host and the pathogen. In most instances, this requires protective sprays.

Resistance refers to the development and use of cultivars that prevent or impede the activity of a pathogen. Although no single cultivar is resistant to all diseases, sources of resistance or tolerance to specific diseases can be used for control.

Therapy is the treatment of plants with something that will inactivate or inhibit the pathogen. In chemotherapy, chemicals are used to inactivate the pathogen. Heat is sometimes used to inactivate or inhibit virus development in infected plant tissues so that newly developing tissue is free of the pathogen.

Part I. Infectious Diseases

Diseases Caused by Fungi

Powdery Mildew

Powdery mildew is probably the most widely distributed and serious disease of glasshouse, garden, and field-grown roses alike. Although the causal fungus was first described in 1819, the disease was present long before then and is now known in all countries in which roses are grown.

Symptoms

Early symptoms are slightly raised, blisterlike, often red areas on the upper leaf surface. The white growth of the fungus, consisting of mycelium and conidiophores, appears as discrete patches on the leaf surfaces of young leaves, which become twisted and distorted and commonly are completely covered with the powdery white growth (Plate 1). Older leaves may not be distorted, but circular or irregular areas may be covered with growth of the mildew fungus (Plate 2). Mature leaves are not usually infected. When environmental conditions are favorable, the affected leaves may fall prematurely.

Fungal growth may develop first on succulent young stem tissues, especially at the base of thorns (Plate 3). This growth persists when stems mature. On garden cultivars, new spring shoots that develop from dormant buds may become infected by the fungus overwintering in rudimentary leaves or bud scales.

The fungus may also attack the flowers and grow abundantly on the pedicels, sepals, and receptacles, especially when the flower bud is unopened (Plate 4). This infection results in flowers of poor quality.

Severe mildew damage reduces leaf growth, the aesthetic value of plants, photosynthetic efficiency and thereby plant growth, and salability of cut flowers. The number of flowers produced may also be reduced, but this has not been conclusively established.

Causal Organism

Theophrastus gave the first account of powdery mildew on rose around 300 BC, but Wallroth in 1819 first described the fungus causing this disease as *Alphitomorpha pannosa*. It was transferred to the genus *Erysiphe* as *E. pannosa* in 1829 and finally was described and placed in the genus *Sphaerotheca* in 1851. Although the fungus has remained identified as *S. pannosa* (Wallr. ex Fr.) Lév., some authorities recognize a division of this species by Woronichine in 1914 into two varieties, var. *rosae* infecting roses and var. *persicae* infecting peach and almond.

Some investigators believe that *S. humuli* also occurs on roses and that most specimens from North America are *S. humuli*, whereas those from Europe are *S. pannosa*. Blumer in 1967 maintained this concept of two species causing powdery mildew on roses but identified the second fungus as *S. macularis*, which is not considered by Salmon to be distinctly different from *S. humuli*. An extensive survey of fresh and herbarium material from around the world led Coyier to the determination that *S. pannosa* and *S. humuli* are not distinctly different and that powdery mildew on rose in the United States is caused by *S. pannosa* (Wallr. ex Fr.) Lév. var. *rosae* Wor. There is some evidence for biological specialization within *S. pannosa*. For example, *S. pannosa* from peach causes large necrotic lesions on apricot leaves, whereas *S. pannosa* from rose cultivar Dorothy Perkins causes small lesions on apricot leaves. It has been suggested but not conclusively established that there are five races of *S. pannosa* var. *rosae* based on pathogen virulence and host susceptibility.

After primary infections on rose, a secondary thick felt, called the "pannose mycelium," is formed, from which the species gets its name. Ascocarps may be found within this mycelium on some cultivars and are variously termed cleistocarps, cleistothecia, or perithecia. Ascocarps are globose to pyriform and are 85–120 μm in diameter, with a few short, pale brown, mycelioid, septate appendages. Asci are broadly oblong to globose, measure 88–115 μm, and contain eight ascospores, each 20–27 × 12–15 μm.

Ascocarps are formed somewhat erratically. In one area, they may form on some cultivars and not on others, and in other areas, they may not form consistently on any one cultivar. There is evidence now that *S. pannosa* var. *rosae* is heterothallic. It is speculated that ascocarps, when formed, may provide a means of overwintering besides that of mycelium in dormant buds; however, little experimental evidence supports this conjecture, and investigators have been unable to get ascospores to germinate.

Disease Cycle

At 20°C and near 100% relative humidity, conidia begin to germinate two to four hours after being deposited on the leaf. A short primary germ tube is produced from one side of the conidium (Plate 5), and within six hours, an appressorial initial is formed. From the bottom of the appressorium, a fine penetration tube pierces the cuticle and enters the epidermal cell, where haustorial initials can be detected after 16–20 hours. Further mycelial growth develops on the leaf surface (Plate 6), and additional haustoria are formed in epidermal cells by 20–24 hours.

By 48 hours, conidiophore initials form as swellings on hyphae immediately above nuclei. Initials elongate and become separated from hyphae by septa once daughter nuclei, formed by division, pass into them. Conidia averaging 22.9–28.6 μm long and 13.6–15.8 μm wide develop at the ends of conidiophores (Plate 7). Successive conidia remain attached to each other in chains, giving the characteristic powdery appearance, or they may be broken off and carried to new infection sites by air currents (Fig. 1). Under optimal conditions, conidial chains are produced by 72 hours after initial infection, although five to seven days are normally required.

Conidia show a diurnal cycle of maturation and abstriction that leads to diurnal periodicity in the number of conidia in the air surrounding rose plants. On a rainless day, the number of conidia released increases as relative humidity decreases; numbers reach a peak from midday to early afternoon and decline as conidiophores are depleted of mature conidia.

In roses grown outdoors in regions with severe

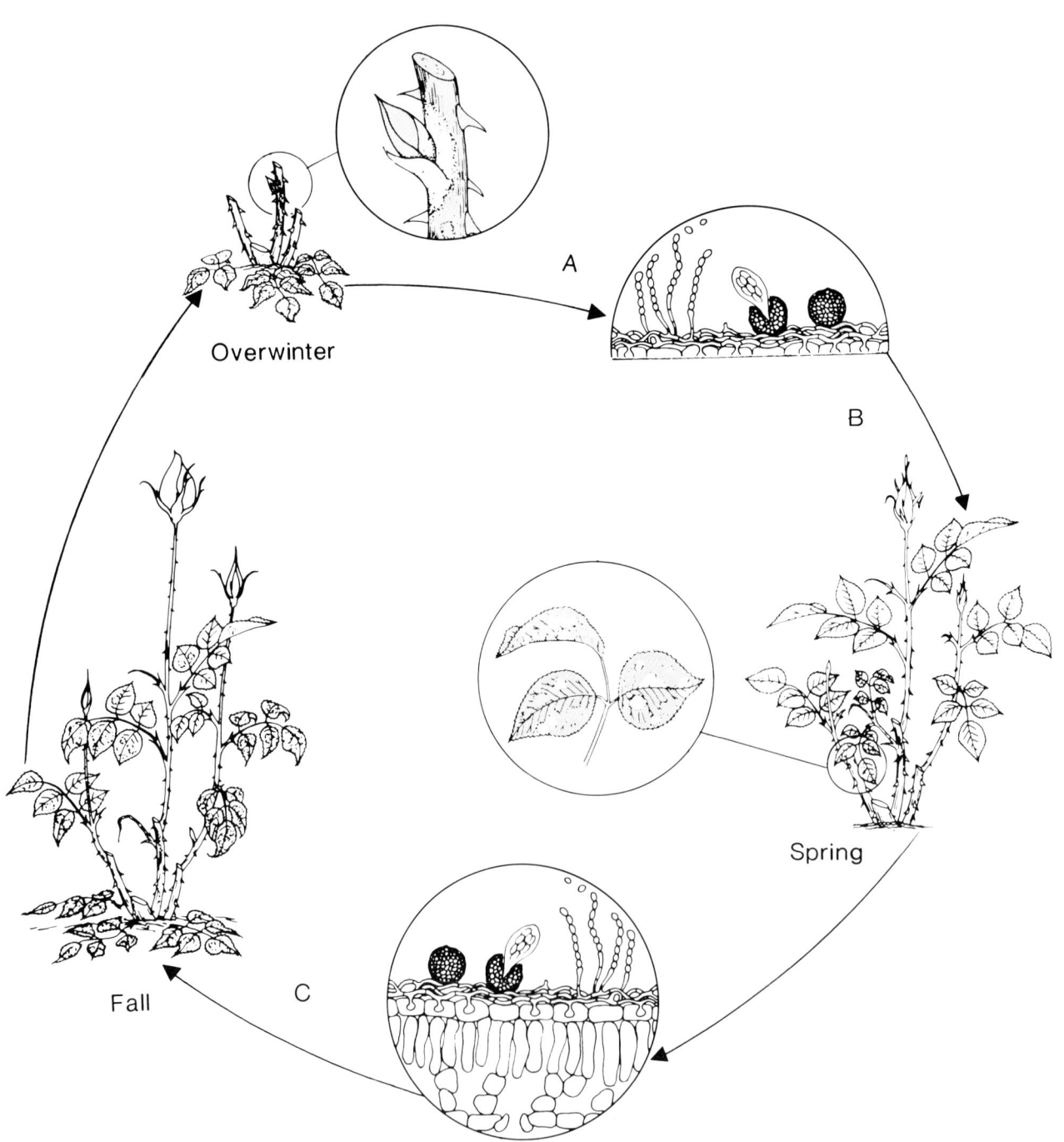

Fig. 1. Powdery mildew disease cycle. *Sphaerotheca pannosa* overwinters in infected canes or buds and in fallen leaves (A). New shoots are infected from overwintering mycelium or from conidia or ascospores produced on fallen leaves (B). Leaves and flowers are infected during the summer by airborne conidia or ascospores produced on infected portions of plants (C).

winters, new infections are initiated from mycelia that overwinter in rudimentary leaves of buds or in the inner bud scales, which survive because they are protected by the outer bud scales. When such buds develop, the resulting shoots become infected and covered with conidia. These conidia are airborne to newly forming leaves, where they initiate a new disease cycle.

The fungus may also overwinter as ascocarps (Fig. 1); however, ascocarps are formed so erratically that this mechanism seems unlikely to be an effective means of overwintering. In mild climates and in glasshouses where growth of roses continues throughout the winter, no survival mechanism is needed, since conidia and new infection cycles are continually produced.

Epidemiology

Differences in the susceptibility of rose cultivars to *S. pannosa* have been reported. Ramblers, climbers, and hybrid teas are generally highly susceptible; wichuraianas are more resistant. Furthermore, the growth stage of the host tissue at the time it becomes infected is important, because the fungus grows well only on young tissues; tissues become more resistant to infection as they age. Typically, mildew development increases as new shoots develop and decreases as these shoots mature and terminate in flower buds. Termination in flower buds is followed by a renewed increase in powdery mildew as lateral buds break and new shoots develop.

In addition to the susceptibility of host tissue, temperature, relative humidity, and the presence of free water also have a strong influence on the growth of *S. pannosa*. The optimal temperature at high humidity is 21°C for conidium germination and 18–25°C for mycelial growth. The optimal relative humidity for germination of conidia is 97–99%. In contrast, powdery mildew development is adversely affected by the presence of films of water on the leaf surface. This effect is most pronounced when leaves are wetted immediately after deposition of conidia. Apparently, conidia do not readily germinate in a free-moisture film. A striking illustration of the influence of liquid water is the fact that during the late 1930s and early 1940s, when spider mites on roses were controlled by frequent water sprays, powdery mildew was rarely a problem, although black spot was very serious. In the last 30 or 40 years, acaricidal sprays and aerosols have replaced this procedure; black spot has essentially disappeared in glasshouse roses, but powdery mildew is again quite serious.

In the field, the most favorable conditions for powdery mildew are as follows. At night, a temperature of 15.5°C and relative humidity of 90–99% allow optimal conidium formation, conidium germination, and infection. Conditions of 26.7°C and 40–70% relative humidity during the day favor the maturation and release of conidia. Several repeated night-day cycles of these conditions are necessary for an epidemic to develop.

Control

New rose cultivars continue to be produced, and many show resistance to powdery mildew. However, few retain a high level of resistance, presumably because of the development of new races of *S. pannosa* that can overcome this resistance. Control has mainly been achieved by protective sprays. In recent years, interest has been increasing in the use of systemic fungicides such as benomyl and triforine.

Control measures for outdoor and glasshouse roses differ somewhat. On outdoor roses, powdery mildew can be expected to occur when rainfall is low or absent, the temperature range is near optimal, and the humidity is high at night and low during the day. Should these conditions occur, protective sprays are necessary. The rapid production of susceptible shoots necessitates repeated application, and the timing of applications is extremely important. Pruning infected shoots at the end of the season and destroying these shoots in regions where winters are severe will help prevent overwintering of the fungus. Raking and destroying fallen leaves from around the bushes at the end of the season may also inhibit overwintering.

On glasshouse roses, powdery mildew can be expected to occur when the temperature range is near optimal and humidity is high at night and low during the day. When these conditions exist, the occurrence of powdery mildew may be forecast three to six days before it appears. Thus, protective fungicidal sprays should be applied and repeated on a seven-day schedule as long as these conditions continue. Other preventive control measures include lowering the night humidity by fans and/or venting or by heating and venting.

Selected References

Bender, C. L. 1982. Pathogenic specialization and heterothallism in *Sphaerotheca pannosa* var. *rosae*. Ph.D. thesis. Oregon State University. 93 pp.

Bender, C. L., and Coyier, D. L. 1982. Identification of five races of *Sphaerotheca pannosa* var. *rosae*. (Abstr.) Phytopathology 72:983.

Coyier, D. L. 1961. Biology and control of rose powdery mildew. Ph.D. thesis. University of Wisconsin. 118 pp.

Pady, S. M. 1972. Spore release in powdery mildews. Phytopathology 62:1099-1100.

Perera, R. G., and Wheeler, B. E. J. 1975. Effect of water droplets on the development of *Sphaerotheca pannosa* on rose leaves. Trans. Br. Mycol. Soc. 64:313-319.

Salmon, E. S. 1900. A monograph of the Erysiphaceae. Mem. Torrey Bot. Club 9. 292 pp.

Tammen, J. F., and Dimock, A. W. 1969. Powdery mildew of roses. Pages 163-171 in: Roses: A Manual on the Culture, Management, Diseases, Insects, Economics and Breeding of Greenhouse Roses. J. W. Mastalerz and R. W. Langhans, eds. Penn. Flower Growers, N.Y. State Flower Growers Assoc., Inc., and Roses Inc. 331 pp.

Wheeler, B. E. J. 1978. Powdery mildews of ornamentals. Pages 411-445 in: The Powdery Mildews. D. M. Spencer, ed. Academic Press, London.

Woronichine, N. 1914. Quelques remarques sur le champignon du blanc de pecher. Bull. Soc. Mycol. Fr. 30:391-401.

Black Spot

Black spot disease has also been called leaf blotch, leaf spot, blotch, rose Actinonema, rose leaf Asteroma, and star sooty mold. It is the most important disease of roses all over the world. Black spot is a minor problem on glasshouse roses, because greater care is taken to avoid syringing plants with water for spider mite control and because humidity is regulated very carefully. In outdoor roses, however, this disease is generally present, frequently epidemic, and a major problem.

Distribution and Occurrence

Black spot is widespread in Europe. It was first reported in Sweden in 1815 and was reported in France, Belgium, Germany, England, and The Netherlands by 1844. The disease is found generally throughout the United States. It was first recorded in the northeastern United States in 1830. Black spot was reported in South America in 1880, in Australia in 1892, in the Soviet Union in 1907, in China in 1910, in Canada in 1911, in Japan in 1914, in Africa in 1920–1922, in India in 1941, and in Turkey in 1947. The pathogen has been widely distributed with cultivated roses and is now generally found even in oceanic islands such as the Philippines, Malta, Hawaii, and New Zealand.

Symptoms

Characteristic black spots 2–12 mm in diameter develop on upper leaf surfaces. These leaf spots are circular or irregularly coalescent with characteristic feathery, radiate, fibrillose margins of subcuticular mycelial strands (Plate 8). Small black acervuli are often visible on the surface (Plates 9 and 10) and may be distributed irregularly or in concentric circles. Conidia may be visible as white, slimy masses on the acervuli. Leaf tissue surrounding the spots turns yellow, and chlorosis extends throughout the leaflet until abscission occurs. The pathogen is actually present only in the lesion itself; the yellow tissue is caused by pathogen metabolites. The yellow tissue exhibits high metabolic activity that is expressed by accumulation of total phenolics and ortho-dihydroxyphenols and amino acids as well as by high enzyme activity. Spots enlarge slowly, taking several weeks to reach 12 mm in diameter. In resistant cultivars or under unfavorable environmental conditions, only tiny black flecks may form and leaves may not turn yellow or abscise.

Yellowing (Plate 11) and abscission of leaflets are associated with ethylene. Leaves with black spot produce large quantities of ethylene; production decreases as leaves become yellow and ceases when leaves turn brown. Infected leaves contain less auxin than healthy ones. The pathogen degrades this abscission-retarding material, thereby hastening leaf abscission.

Raised, purple-red, irregular blotches develop on the immature wood of first-year canes of susceptible cultivars (Fig. 2). Spots later become blackened and blistered; they contain acervuli but lack fibrillose strands. Lesions are often small and rarely kill branches but are extremely important in the survival of the pathogen over the winter.

Petioles and stipules may have inconspicuous black spots similar to those found on leaves. Petioles may be girdled without abscising. Peduncles, fruit, and sepals may have similar symptoms. Petals may have red flecks accompanied by moderate distortion. Acervuli frequently occur in the lesions.

Causal Organism

Marssonina rosae (Lib.) Lind (*Asteroma rosae, Actinonema rosae, Marsonia rosae*), the imperfect stage of the black spot pathogen, was described in France in 1827. The perfect stage, *Diplocarpon rosae* Wolf, was described in New York in 1912. The pathogen is quite host-specific and approaches obligate parasitism. Pathogenic races of the fungus are reported.

Parasitic mycelia of *M. rosae* are characteristically subcuticular, radiate, branching, and single or in strands of parallel hyphae. Hyphae are hyaline when young but darken with age. Haustoria are formed in host cells. Acervuli are subcuticular and irregularly rupture the cuticle. Acervuli vary in diameter from 50 to 400 μm. Each acervulus bears two-celled, hyaline conidia (Fig. 3B). The conidia (15–25 × 5–7 μm) are smooth with a sticky surface and occur in a white, slimy mass.

Dead leaf tissues contain intercellular and intracellular mycelia and lack haustoria. Apothecia are rarely formed; they have been reported twice from the northern United States and Canada (in the period from October to December) and twice from England (in the period from April to May). Apothecia measure 100–250 μm in diameter and have a circular subcuticular shield of dark brown, thick-walled cells. Asci (70–80 × 15 μm)

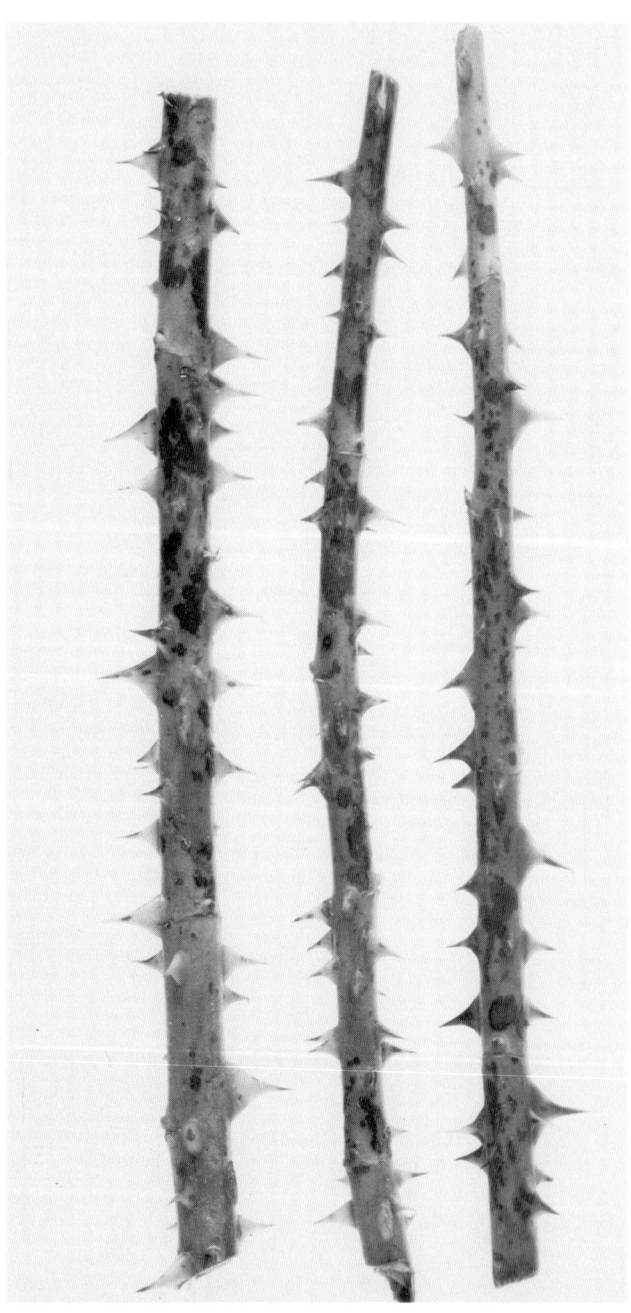

Fig. 2. Canes infected with the black spot fungus (*Diplocarpon rosae*).

contain eight hyaline ascospores (20–25 × 5–6 μm). Ascospores are forcibly discharged and are airborne; they are not water-dispersed.

The fungus grows on potato-dextrose or malt agar but requires 15–37 days to form a visible colony from a single spore and one month to reach a colony diameter of 2–9 mm. Virulence may be lost in a few months in culture.

Disease Cycle

Leaves are most susceptible while still expanding (6–14 days old). To germinate, conidia must be wetted at least five minutes even if held at 100% humidity. Regardless of the relative humidity, conidia must be immersed in water and must be continuously wet for at least seven hours for any infections to occur. A conidium germinates in 9–18 hours on a moist leaf at 22–26°C. Germ tubes emerge from the larger cell of the conidium and infrequently from both cells. Germ tubes penetrate the leaf cuticle and grow between the cuticle and epidermal cells. Fine hyphae branch vertically, penetrate the epidermal cell wall, expand into the bulbous structure outside the plasmalemma, and produce haustoria. Haustoria may form within host cells 15 hours after infection. Secondary mycelium forms on the second day, and in three to five days, parallel subcuticular strands are formed. By the sixth and seventh days, there may be as many as five to eight haustoria per cell.

Infections on lower leaf surfaces resemble those on

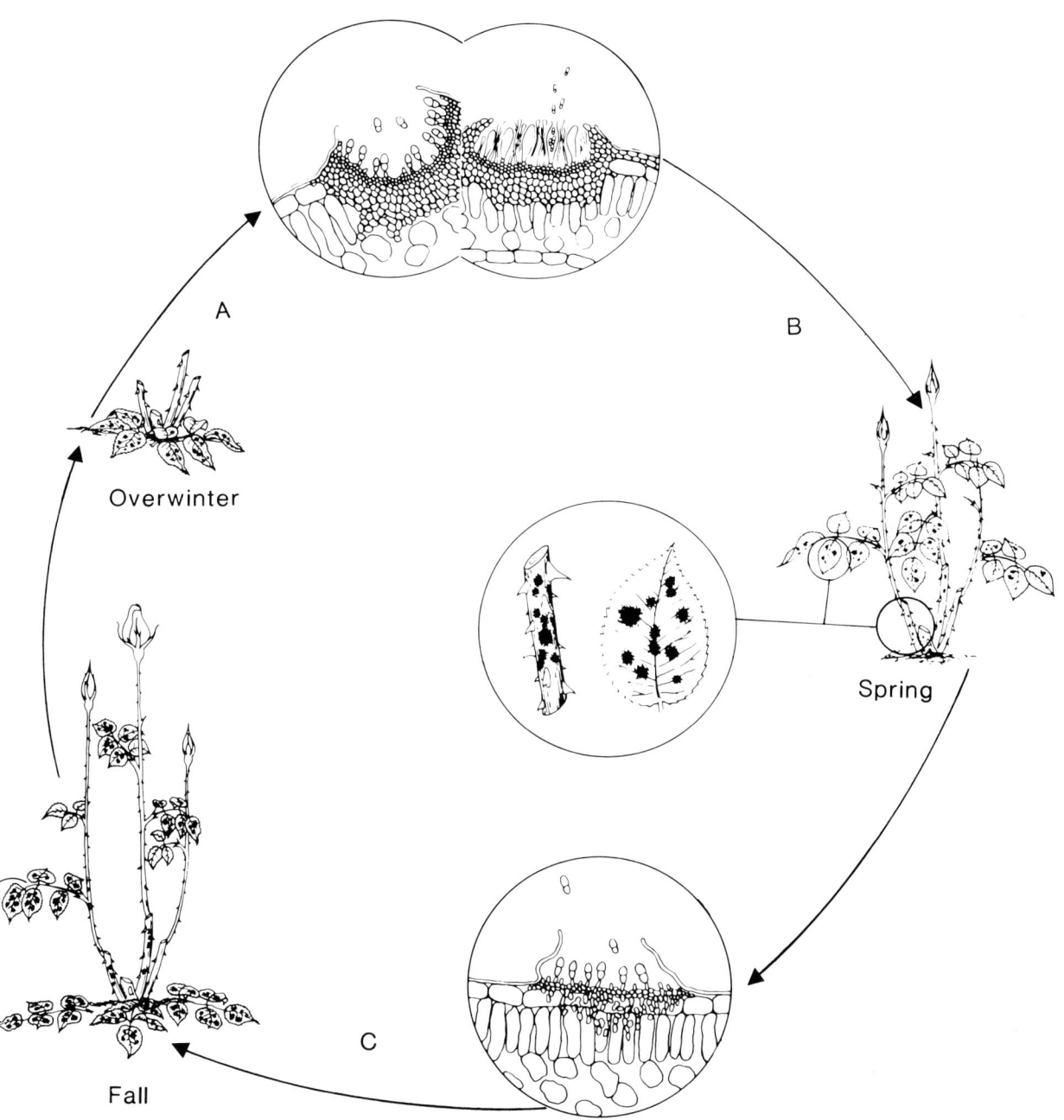

Fig. 3. Black spot disease cycle. *Diplocarpon rosae* overwinters in infected canes or buds and in fallen leaves (A). New shoots are infected from overwintering mycelium or from water-splashed conidia or airborne ascospores produced on fallen leaves, sometimes in the same lesion (B). Leaves and canes are infected during summer by airborne ascospores or water-splashed conidia produced on infected leaves (C).

upper surfaces, except that hyphae soon grow through the mesophyll, developing on both upper and lower surfaces. Symptoms appear in 3–16 days, depending on temperature and inoculum.

Acervuli form in 11 days on the upper leaf surface and one month on the lower. When the relative humidity drops at the location of the enlarging acervulus, the cuticle breaks, exposing a slimy mass of conidia. Conidia may be produced 10–18 days after infection. Production of conidia by a given acervulus decreases after about a week and may end after 10 days, but new acervuli are continually formed at the margins of spots.

Some conidia adhere to ruptured cuticle fragments, but most are dispersed in rain water or condensate. Conidia are disseminated by splashing water, by people during cultivation, or by contact with sticky body parts of insects. Fallen leaves blown by the wind may disperse the pathogen locally, but conidia are usually airborne only in water drops. Ascospores are found so rarely that they are unimportant in dispersal. Infected plants may introduce the fungus into glasshouses.

The fungus does not survive in the soil, and conidia adhering to tools, benches, etc., remain viable no longer than one month. In areas with mild climates and in glasshouses, the fungus remains active on the host throughout the year. The fungus overwinters as mycelia in fallen leaves or in infected canes (Fig. 3). Overwintering in fallen leaves may be as conidia in existing acervuli, as acervuli in which new conidia are produced in spring, or as mycelia that in spring produce either new acervuli or apothecia in which conidia and ascospores form.

Epidemiology

M. rosae tolerates a wide range of temperatures (15–27°C), mainly through an inverse relation of humidity and moisture. Conidium germination is optimal at 18°C; at this temperature, germination begins in nine hours and may reach 96% in 36 hours. The pathogen is sensitive to high temperatures. Conidia are killed without germinating at 33°C, whereas at 30°C, they may germinate but fail to develop further. Mycelial growth is optimal at 21°C and is halted after eight weeks at 33°C.

Conidial infection of leaves is optimal at 19–21°C, and symptoms may develop within three to four days at 22–30°C. Disease development is optimal at 24°C. No infections occur if leaf surfaces are dried within seven hours of inoculation and incubated at 15–24°C. The success of infections increases if leaves remain moist for 24 hours before drying. No infections occur in dry air. Even at 100% relative humidity, no germination occurs if conidia have not been wetted, and at relative humidity below 90%, mature conidia must be immersed in water for at least seven hours before they can cause infection. Conidia begin to germinate as early as eight hours after being wetted.

Good air circulation around bushes in glasshouses or in the field hastens drying and reduces black spot. However, cool moist ocean winds in coastal regions or overhead sprinkling favors infection. Disease development is restricted in arid regions or in glasshouses with low humidity where sprinkled plants dry rapidly. Summer heat and winter cold limit development of epidemics in rainy regions.

Control

Leaves should not be allowed to remain wet or at very high humidity for more than 7–12 hours. Plants should not be syringed with water; if syringing is practiced, it should be done only on bright mornings with rising temperatures. Excessive watering should be avoided during dark, humid weather.

Removing leaves from the ground and pruning canes that contain lesions will reduce overwintering of the pathogen. Dense planting should be avoided to allow good air circulation through the leaf canopy.

Fungicidal sprays should be used during periods of the year when conditions are favorable for black spot development. In the northeastern United States, preventive sprays should be initiated around February 15 to March 1 and should continue bimonthly until leaf emergence on outdoor-grown roses. Thereafter, fungicides should be applied every seven days. Good control can be obtained with fungicidal sprays every 14 days if surfactants such as Ex-800, X-77, Triton B-1956, or W-P NCF are used in sprays with wettable powder fungicides.

Black spot resistance in roses is rare, especially in tetraploid roses. Cultivars reported to be highly resistant are David Thompson, Bebe Lune, Coronado, Ernest H. Morse, Fortyniner, Grand Opera, Lucy Cromphorn, Sphinx, Tiara, Carefree Beauty®, and Simplicity. The occurrence of pathogenic races of the fungus makes it difficult to develop resistant cultivars. In general, teas, hybrid teas, hybrid perpetuals, Pernetianas, Austrian briers, and polyanthas are quite susceptible and rugosa hybrids, moss roses, and wichuraianas are more resistant. Among rootstocks, the Welch and Tate strains of *Rosa multiflora* are highly resistant, and *R. odorata*, *R. manetti*, *R. caudata*, IXL, Texas Wax, and Ragged Robin are susceptible; however, it has been reported that disease susceptibility is a characteristic of the scion, and understocks are not known to influence the disease susceptibility of the scion.

Selected References

Aronescu, A. 1934. *Diplocarpon rosae*: From spore germination to haustorium formation. Torrey Bot. Club Bull. 61:291-329.

Baker, K. F., and Dimock, A. W. 1969. Black spot. Pages 172-184 in: Roses: A Manual on the Culture, Management, Diseases, Insects, Economics and Breeding of Greenhouse Roses. J. W. Mastalerz and R. W. Langhans, eds. Penn. Flower Growers, N.Y. State Flower Growers Assoc., Inc., and Roses Inc. 331 pp.

Bhaskaran, R., Purushothaman, D., and Ranganathan, K. 1974. Physiological changes in rose leaves infected by *Diplocarpon rosae*. Phytopathol. Z. 79:231-236.

Bolton, A. T., and Svejda, F. J. 1979. A new race of *Diplocarpon rosae* capable of causing severe black spot on *Rosa rugosa* hybrids. Can. Plant Dis. Surv. 59:38-40.

Castledine, P., and Roberts, A. V. 1981. Cuticular resistance to *Diplocarpon rosae* blackspot fungus disease of roses. Trans. Br. Mycol. Soc. 77:665-666.

Cook, R. T. A. 1981. Overwintering of *Diplocarpon rosae* at Wisley, England. Trans. Br. Mycol. Soc. 77:549-556.

Knight, C., and Wheeler, B. E. J. 1977. The germination of *Diplocarpon rosae* on different rose cultivars. Phytopathol. Z. 91:346-354.

Morrison, L. S., and Russell, C. C. 1976. Timing of fungicide adjuvant mixtures for control of rose black spot. Plant Dis. Rep. 60:634-636.

Palmer, J. G., Sachs, I. B., and Semeniuk, P. 1978. The leaf spot caused by *Marssonina rosae* observed in scanning electron microscopy and light microscopy. Scanning Electron Microsc. 2:1019-1026.

Singh, S. N. 1980. Effect of rootstock on growth, flowering and disease resistance of hybrid tea roses. Prog. Hortic. 12(3):5-14.

Sohi, H. S., and Prakash, O. M. 1974. Reaction of rose varieties against black spot, *Diplocarpon rosae*, disease. Indian Phytopathol. 27:119-120.

Svejda, F. 1979. David Thompson rose, *Rosa rugosa*. Can. J. Plant Sci. 59:1167-1168.

Svejda, F., and Bolton, A. T. 1980. Resistance of rose hybrids to three races of *Diplocarpon rosae* blackspot. Can. J. Plant Pathol. 2:23-25.

Rust

Nine species of the rust fungus *Phragmidium* are found on roses: *P. mucronatum* (Pers.) Schlecht., *P. tuberculatum* Mull., *P. fusiforme* Schrot., *P. rosae-pimpinellifoliae* (Rabh.) Diet., *P. americanum* (Pk.) Diet., *P. montivagum* Arth., *P. rosae-californicae* Diet., *P. rosicola* (Ell. & Ev.) n. comb., and *P. speciosum* (Fr.) Cke. All of these species have been reported on native species of rose, and *P. mucronatum, P. americanum, P. fusiforme, P. speciosum*, and *P. tuberculatum* may occur on cultivated roses as well. *P. mucronatum* is the most common rust species in the United States on hybrid tea roses and other roses with large, firm leaflets. In England, *P. mucronatum, P. tuberculatum, P. fusiforme*, and *P. rosae-pimpinellifoliae* have been found on more than 200 named cultivars of bush rose (hybrid teas and floribundas); *P. tuberculatum* is the most common. Although rust on rose is generally widespread, it is more common in southern and eastern England, the western United States, and other geographic areas where cool temperatures and high moisture during certain times of the year are conducive to disease development.

Symptoms

The disease first appears on leaves and other green parts of the plant as powdery pustules of orange aeciospores usually confined to lower leaf surfaces (Plates 12 and 13). In early spring, spore masses may be inconspicuous and go unnoticed. As pustules develop, they become visible on upper leaf surfaces as orange or brown spots (Plate 14). Young stems and sepals also may become infected (Plate 15) and finally distorted.

Cultivars vary widely in susceptibility and reaction to infection. Leaves of some cultivars may become covered with pustules and yet remain attached to the plant, while a single rust pustule on a leaflet of another cultivar will cause the leaflet to fall. Susceptible and moderately susceptible cultivars are Arlene Francis, Aztec, Baby Blaze, Betsy McCall, Blue Moon, Buccaneer, Christopher Stone, Chrysler Imperial, Circus, Confidence, Dearest, Elizabeth of Glamis, Embers, Fragrant Cloud, Fusilier, Golden Girl, Golden Masterpiece, Heat Wave, Helen Traubel, Jeanie, Josephine Bruce, Kordes Perfecta, Montezuma, New Yorker, Nocturne, Peace, Piccadilly, Pink Peace, Pink Radiance, Queen Elizabeth, Siren, Spartan, Sutter's Gold, Talisman, The Doctor, Virgo, Vogue, Wendy Cussons, White Bouquet, White Knight, and White Swan. Many new cultivars not included in this list are also susceptible.

The summer uredial stage has reddish orange pustules and may repeat every 10–14 days under favorable environmental conditions. The repeating summer uredial stage is followed by wilting and defoliation of susceptible cultivars. In mild climates, the uredial stage continues; in cooler areas, black, telial-stage pustules are formed toward autumn (Plate 16).

Causal Organisms

P. mucronatum and *P. tuberculatum* are the two species of *Phragmidium* most commonly found to cause rust on cultivated rose. *P. mucronatum* was the first fungal parasite to be seen with the microscope in 1665 by Hooke, who gave a careful drawing, complete with scale, of the teliospore on a rose leaf.

Rust fungi with a complete life cycle have five different spore forms, which are numbered 0 to IV: 0, spermatia formed in spermagonia; I, aeciospores in aecia; II, urediospores in uredia; III, teliospores in telia; and IV, basidiospores in basidia. In heteroecious rusts, spore stages 0 and I are formed on one host and II and III on another. Stage IV always follows III after germination. Although most autoecious rusts have all spore forms on one host, a few short-cycle rusts fail to produce all the spore stages.

The nine species of *Phragmidium* that cause rust on rose are listed below, along with the spore stages found on roses. Species are determined by the number of cells in the teliospores (Plate 17).

P. mucronatum: 0, I on leaves and stems; II, III on leaves of cultivated roses. Teliospores have five to nine cells.

P. tuberculatum: 0, I on leaves and stems; II, III on leaves of cultivated roses. Teliospores have three to five cells.

P. americanum: 0, I, II, III on leaves of native and cultivated roses. Teliospores have 8–11 cells.

P. fusiforme: 0, I, II, III on leaves of native and cultivated roses but not widely distributed. Teliospores have 5–11 cells.

P. montivagum: 0, I, II, III on leaves of many species of roses. Teliospores have six to nine cells.

P. rosae-californicae: 0, I, II, III on leaves of many species of roses. Teliospores have 8–11 cells.

P. rosicola: III on *Rosa engelmanii* and *R. suffulta*. Teliospores are one-celled and nearly round.

P. speciosum: 0, I on stems and leaves; III on stems of native and cultivated roses. Teliospores have four to eight cells.

P. rosae-pimpinellifoliae: 0, I on stems; II and III on leaves; on *R. foetida, R. spinosissima*, and hybrid cultivar Agnes. Teliospores have five to seven cells.

Axenic cultures of *P. mucronatum* can be grown successfully on agar media containing yeast extract, peptone, and casein hydrolysate thickly seeded with urediospores.

Epidemiology

Spores from rust pustules are air-transmitted and infect rose leaves through stomatal openings. The optimal temperatures for development of the disease are 18–21°C, and continuous moisture for two to four hours is essential for establishment of infection. On susceptible cultivars in glasshouses, infection is likely to be severe near ventilators where condensation occurs.

Teliospores have a stalked pedicel, which has a swelling in the lower portion (Plate 17). A gelatinized wall forms at the base of the teliospore and may fix the spore in a position suitable for producing basidia in the spring.

In late summer and early autumn, black pustules appear on outdoor-grown roses, often in the same affected leaf areas in which teliospores are found (Plate 18). Black pustules overwinter within the leaf and stem tissues (Plate 19) after leaves fall and later produce spores for spring infections. In areas with severe winters, the rust fungus may not overwinter, and the summers are usually dry enough that there is little increase in the rust disease. High summer temperatures inhibit infection; urediospores retain viability for only a week at 27°C.

Control

Removing infected leaves during the season and all old leaves left at the time of winter or early spring pruning before new leaves appear helps to reduce inoculum levels and prevent early appearance of disease. Spring pruning of old canes will help to eliminate rust carry-over on canes.

Any means of preventing condensation in glasshouses will aid in controlling rust, since free moisture for several hours is essential for infection.

Preventive fungicidal sprays should be applied every seven days during periods when environmental conditions favor disease development.

Selected References

Bhatti, M. H. R., and Shattock, R. C. 1980. Axenic culture of *Phragmidium mucronatum*. Trans. Br. Mycol. Soc. 74:595-600.

Horst, R. K. 1979. *Phragmidium*. Pages 369-370 in: Westcott's Plant Disease Handbook. 4th ed. Revised by R. K. Horst. Van Nostrand Reinhold Company, New York. 803 pp.

Howden, J. C. W., and Jacobs, L. 1973. Report on the rust work at Bath. The Rose Annual 1973:113-119.

Ingold, C. T., Davey, R. A., and Wakley, G. 1981. The teliospore pedicel of *Phragmidium mucronatum*. Trans. Br. Mycol. Soc. 77:439-442.

Nichols, L. P., and Nelson, P. E. 1969. Foliage diseases. Pages 185-187 in: Roses: A Manual on the Culture, Management, Diseases, Insects, Economics and Breeding of Greenhouse Roses. J. W. Mastalerz and R. W. Langhans, eds. Penn. Flower Growers, N.Y. State Flower Growers Assoc., Inc., and Roses Inc. 331 pp.

Verticillium Wilt

Verticillium wilt was first reported in the eastern United States in 1924 and then in the western United States two years later. The disease has now been reported in other parts of the United States and in Canada, Europe, South America, and New Zealand. Because the causal fungus has a wide host range, including other ornamental plants as well as fruit and vegetable crops, it is thought to occur on roses wherever they are grown.

Verticillium wilt is troublesome on glasshouse-grown roses and on roses grown outdoors for greenhouse plantings and cut flowers. However, the disease may also occur on garden roses. Field-grown roses planted in fields previously cropped to field or vegetable crops that are susceptible to Verticillium wilt are particularly prone to the disease, because the pathogen may be present in the soils.

Symptoms

The characteristic initial symptoms of Verticillium wilt are wilting leaves at the tips of young canes and a yellowing of lower leaves. After a few days, permanent wilting occurs, and the leaves generally turn yellow (Plates 20–22) and finally brown as they wither and die. Defoliation progresses from the base of canes upward. Canes that show symptoms may continue to grow normally in subsequent seasons, or they may die back. Dieback usually begins at the tip and progresses downward, and necrotic lesions or purple-black streaks frequently occur along the length of the shoot (Plates 23 and 24). Progressive dieback can result in death of entire plants. Vascular discoloration, common on other hosts, is not usually evident in infected roses. Symptoms may be confused with rose wilt, which is thought to be caused by a virus (see Rose Wilt).

Symptoms appear during periods of stress, such as drought in midsummer or later. In many instances, leaves may wilt during the day and recover at night. Symptoms on outdoor-grown roses are usually less severe than those on glasshouse-grown roses. Field-grown roses recover naturally, and infection periods are limited primarily to winter and spring. This seasonal effect may be caused by the cool-temperature requirements of the pathogen. Seasonal variation is less evident in glasshouses, where the pathogen may be active throughout the history of the planting. Tissues formed on roses grown in glasshouses are more succulent than those on roses grown outdoors and thus may be more susceptible to rapid infection by *Verticillium*.

A potential danger is that infected plants may tolerate infection and show no symptoms when growing conditions are favorable to the crop. *Verticillium*, however, may be readily isolated from infected tissues even if no symptoms are evident. Infected but symptomless plants shipped by a supplier often show symptoms after being planted by the grower.

Causal Organisms

Verticillium wilt is caused by *Verticillium albo-atrum* Reinke & Berth. or by *V. dahliae* Kleb. Conidia are one-celled, hyaline, and globose to ellipsoid. They are formed at the tips of whorled branches of conidiophores. Conidia separate readily from the conidiophore tips and can be found in vascular elements of infected plants (Plate 25). *V. dahliae* forms microsclerotia; *V. albo-atrum* does not.

Control

Because the causal organisms are soil inhabitants, they are frequently disseminated in infested soils. Therefore, soils used for field- and glasshouse-grown roses should be sterilized or fumigated before planting.

Rootstocks used in the field production of rose plants vary in susceptibility to *Verticillium*. *R. odorata* and Ragged Robin are susceptible, whereas *R. multiflora* and Dr. Huey are somewhat resistant and *R. manetti* is very resistant. Although *Verticillium* has been transmitted through buds used in grafting, the significance of bud transmission is negligible.

Verticillium wilt is very difficult to control after it appears in glasshouse-grown roses used for cut flowers. Thus, effective control measures must begin before the disease is found. It is extremely important for producers of field-grown roses to control Verticillium wilt.

Selected References

Hammett, K. R. W. 1971. Symptom differences between rose wilt virus and Verticillium wilt of roses. Plant Dis. Rep. 55:916-920.

Horst, R. K. 1979. *Verticillium.* Pages 439-443 in: Westcott's Plant Disease Handbook. 4th ed. Revised by R. K. Horst. Van Nostrand Reinhold Company, New York. 803 pp.

Nelson, P. E., and Nichols, L. P. 1969. Root and stem diseases. Pages 203-210 in: Roses: A Manual on the Culture, Management, Diseases, Insects, Economics and Breeding of Greenhouse Roses. J. W. Mastalerz and R. W. Langhans, eds. Penn. Flower Growers, N.Y. State Flower Growers Assoc., Inc., and Roses Inc. 331 pp.

Raabe, R. D., and Wilhelm, S. 1955. Field studies of rose rootstock infections by the Verticillium wilt fungus. (Abstr.) Phytopathology 45:695.

Visser, S., and Kotze, J. M. 1975. Reactions of various crops to inoculation with *Verticillium dahliae* isolated from wilted strawberry plants. Phytophylactica 7:25-30.

Downy Mildew

Downy mildew of roses was first reported in 1862 in England and from then into the early 1900s was reported throughout Europe from France through the Scandinavian countries to the Soviet Union. The disease was first reported in the midwestern United States in 1880 and more recently from all parts of the country. Downy mildew has occurred on roses in British Columbia in Canada and has also been reported in Iceland. Although most of the literature records downy mildew as a disease of roses in areas north of the Tropic of Cancer, it has been reported in Brazil and its geographic distribution has expanded in recent years. The disease is known to occur on roses in Colombia, Mauritius, Israel, and Egypt, and it is widespread in Australia. All rose cultivars are susceptible, although disease severity is variable. Wild *Rosa* spp., including *R. californica, R. centifolia, R. canina, R. rubiginosa,* and *R. indica,* are also susceptible.

Symptoms

Symptoms of downy mildew occur on leaves, stems, peduncles, calyxes, and petals. Infection is generally restricted to young, apical plant growth. Leaves develop purplish red to dark brown irregular spots (Plates 26 and 27), and leaflets may turn yellow; islands of normal green tissue up to 1 cm in diameter are often observed on the yellow leaflets. Leaf abscission may be severe. Foliar symptoms may resemble burns from pesticide toxicity (Plate 28).

Under humid, cool conditions, conidia and conidiophores appear copiously on the lower surfaces of leaves (Plate 29), but under less favorable conditions, spore production is sparse (hence the species name of the causal agent, *Peronospora sparsa*) and difficult to detect. Purplish to black areas varying from small spots to areas 2 cm or more in length appear on stems and peduncles. Similar spots and dead tips develop on calyxes, and infected twigs may be killed.

Downy mildew differs from powdery mildew of rose in that the grayish spores of downy mildew are produced only on lower leaf surfaces; powdery mildew is external, and the superficial, white mycelia and spores are found on both leaf surfaces. Downy mildew also causes spectacular foliar abscission, unlike powdery mildew, which kills tissue only when infection is severe over an extended period.

Causal Organism

Downy mildew is caused by *Peronospora sparsa* Berk. Mycelia are intercellular in host tissues. Erect sporangiophores on lower leaf surfaces are dichotomously branched, with sporangia borne on sharply pointed terminal branches (Fig. 4). Under humid conditions, conidiophores are about 350 μm long; subelliptical sporangia 17–22 × 14–18 μm are produced abundantly through stomata on the lower leaf surface.

Oospores may be found in infected leaves, sepals,

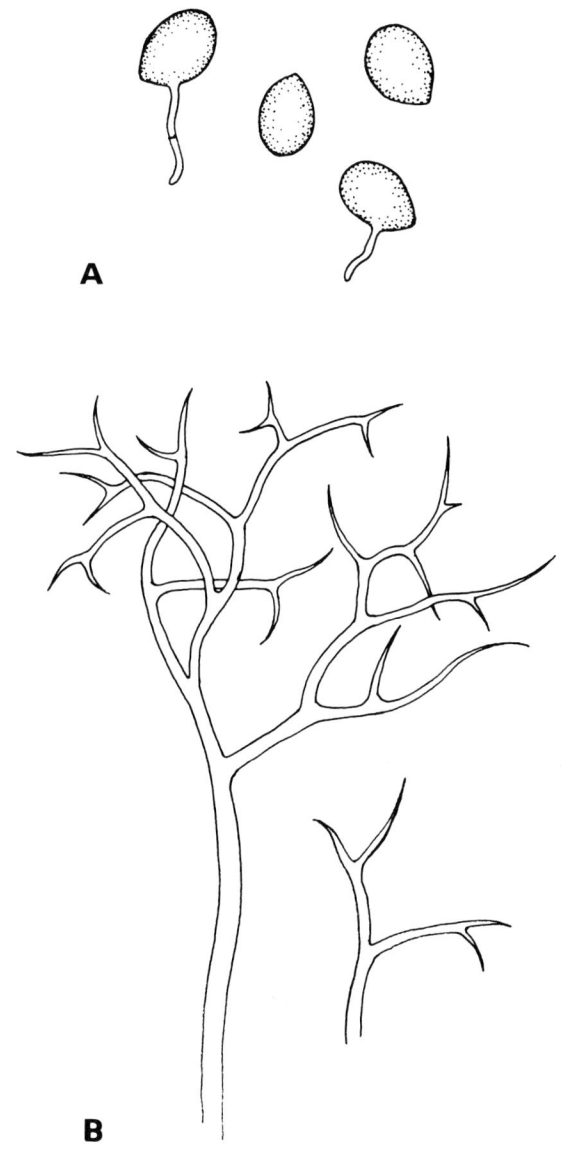

Fig. 4. Sporangia **(A)** and sporangiophores **(B)** of *Peronospora sparsa*.

flower buds, and stems. The fungus may overwinter in stems as dormant mycelia without oospores. Sporangia may be produced for long periods of time when high humidity and cool temperatures persist. Roses are unaffected by downy mildew when the humidity is less than 85%. The optimal temperature for spore germination is 18°C. Spores do not germinate at 5°C and are killed by exposure to 27°C for 24 hours.

Sporangia germinate within four hours in water, and sporulation on leaf surfaces may occur in three days under ideal conditions. Spores may survive and be viable on dried fallen leaves for as long as one month.

Control

Lowering the humidity by ventilation and aeration and/or raising the temperature to 27°C during the warmer parts of the day and evening is helpful for controlling downy mildew in glasshouse rose production. Special care should be taken to prevent a sudden slight drop in the temperature at sunset, which greatly increases relative humidity. The humidity should not remain above 85% for more than three hours. Gutter leaks and wet floors, benches, and walks should be avoided in glasshouses.

Although downy mildew is most severe on glasshouse-grown roses, it is also a serious problem on field-grown roses when environmental conditions favor disease development. Preventive fungicidal sprays should be used when disease conditions are ideal for infection.

Sanitation is important to prevent seasonal carryover of the pathogen. Infected leaves, stems, and flowers should be destroyed. Because the pathogen may overwinter in stems as dormant mycelia and oospores, growers should carefully remove cuttings and plant parts suspected of carrying the downy mildew fungus in characteristic symptomatic tissues.

Selected References

Baker, K. F. 1953. Recent epidemics of downy mildew of rose. Plant Dis. Rep. 37:331-339.

Gill, D. L. 1977. Downy mildew of roses in Georgia. Plant Dis. Rep. 61:230-231.

Nichols, L. P., and Nelson, P. E. 1969. Foliage diseases. Pages 185-195 in: Roses: A Manual on the Culture, Management, Diseases, Insects, Economics and Breeding of Greenhouse Roses. J. W. Mastalerz and R. W. Langhans, eds. Penn. Flower Growers, N.Y. State Flower Growers Assoc., Inc., and Roses Inc. 331 pp.

Stahl, M. 1973. Some observations on downy mildew on roses. Nachrichtenbl. Dtsch. Pflanzenschutzdienstes (Braunschweig) 25:161-162.

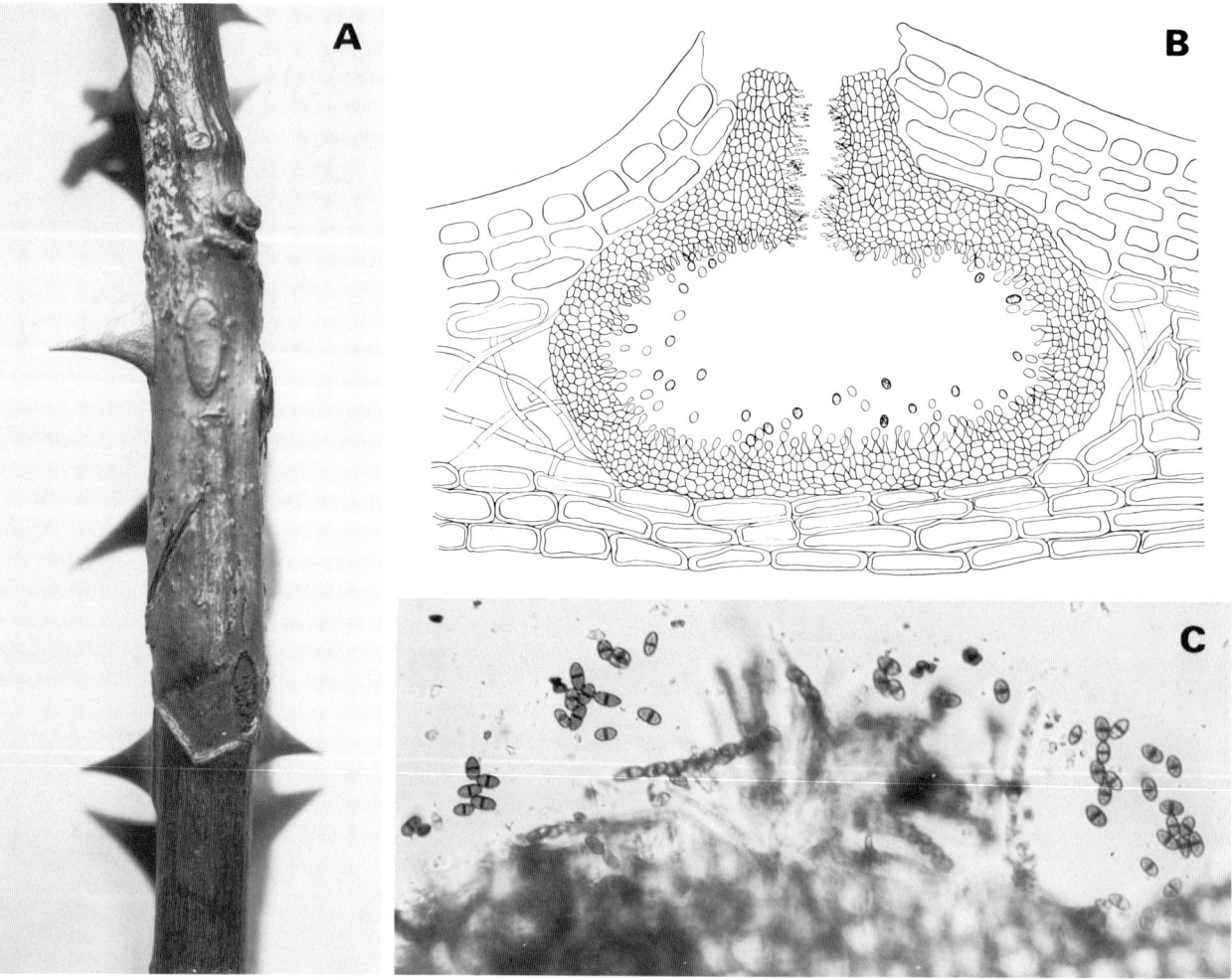

Fig. 5. Brand canker caused by *Coniothyrium wernsdorffiae*. **A,** Canker on cane with small, black, elevated pycnidial spots. **B,** Pycnidium developing in bark of cane, which is slit as pycnidium enlarges. **C,** Conidia of *C. wernsdorffiae*.

Brand Canker

Brand canker was first reported on roses in Europe in the late 1800s, and the fungus pathogen was named in Germany in 1905. At that time, the disease was widespread in Europe. It was first recognized in the United States in 1925 and has also been reported in Canada and the Soviet Union. Brand canker occurs primarily on outdoor-grown roses.

Symptoms

Characteristic symptoms are confined to rose canes and vary greatly in size. Initial symptoms are small, dark reddish spots that gradually enlarge and acquire a definite reddish brown or purple margin that contrasts sharply with the green of the rose cane. The center of the canker turns light brown as the cells die, and pycnidia of the fungus appear as small, black, elevated spots in the canker (Fig. 5A). As the pycnidia increase in size, small longitudinal slits appear in the epidermis, exposing masses of spores (Fig. 5B). Cankers formed under the winter protection of snow are black when roses are first uncovered in the spring; thus the name brand canker or *Brandfleckenkrankheit*, which means fire-spot disease. After one or two weeks in the light and air, canker symptoms assume the characteristics described above. Cankers may girdle rose canes, resulting in dieback of growing tips and eventual death of the entire cane. The disease is sometimes confused with common canker.

Causal Organism

Coniothyrium wernsdorffiae Laub. is the fungus that causes brand canker. Pycnidia are black, globose, separate, erumpent, and ostiolate. Conidia are small, one-celled, and ovoid or ellipsoid (Fig. 5C). The olive brown conidia are nearly twice the size of those of *C. fuckelii* and are released through longitudinal slits in the epidermis of rose canes rather than spread in a sooty mass under the epidermis as are conidia of *C. rosarum*.

C. wernsdorffiae is a cold-temperature fungus that infects rose canes under winter covering. Infection occurs through insect wounds, thorn scars, and scratches, and occasionally through dormant buds.

Control

Winter protection that maintains moist conditions around rose canes should be avoided. Care should be taken to avoid injury to rose canes. Pruning cuts should be made immediately above the node without actually cutting the nodal tissue and at an angle to leave the minimum stub of dead wood. Cuts at nodes will callus normally and prevent infection by *C. wernsdorffiae*. When long stubs are left, the fungus becomes established in the dead wood of the stub, develops down the stem and through the node, and may kill the entire stem or plant.

Infected canes should be removed at any node below the visibly diseased area. Sharp pruning tools should be used to obtain clean cuts and avoid wounding. Fungicidal sprays may be used to cover wounds.

Selected References

Horst, R. K. 1979. *Coniothyrium*. Pages 157-158 in: Westcott's Plant Disease Handbook. 4th ed. Revised by R. K. Horst. Van Nostrand Reinhold Company, New York. 803 pp.

Nelson, P. E., and Nichols, L. P. 1969. Root and stem diseases. Pages 203-210 in: Roses: A Manual on the Culture, Management, Diseases, Insects, Economics and Breeding of Greenhouse Roses. J. W. Mastalerz and R. W. Langhans, eds. Penn. Flower Growers, N.Y. State Flower Growers Assoc., Inc., and Roses Inc. 331 pp.

Protsenko, E. P., and Chelyshkina, B. A. 1973. Rose brand canker, *Coniothyrium wernsdorffiae*. Mikol. Fitopatol. 7:119-124.

Waterman, A. M. 1930. Diseases of rose caused by species of *Coniothyrium* in the United States. J. Agric. Res. 40:805-827.

Common Canker (Rose Graft Canker)

Common stem canker was first reported on roses in Europe in the late 1800s, and the fungus pathogen was named in Germany in 1905. At that time, the disease was widespread in Europe; it was first recognized in the United States in 1925. Common canker occurs on both outdoor and glasshouse-grown roses and probably is widespread in occurrence.

Symptoms

Wounds are necessary for infection. Cankers begin as small yellow to red spots in the bark and gradually expand. The centers of the cankers become light brown and the margin a darker brown (Fig. 6). Epidermal tissue within the canker dries out and shrinks, sometimes cracking, exposing masses of small conidia (Fig. 6B-D and K-Q). Cankers may girdle the stem, causing wilting and death of the plant parts above the canker. Small pycnidia are sometimes produced in abundance on the canker. Symptoms occur initially at the union of stocks and scions in warm, moist propagating areas, and development continues in the dead wood when plants are moved into the glasshouse. The disease can be serious on rose plants in storage and on recently planted roses, especially if they are placed under stress.

Causal Organism

Coniothyrium fuckelii Sacc., the imperfect stage of *Leptosphaeria coniothyrium* (Fckl.) Sacc., causes common canker of rose. It is thought to be synonymous with *C. rosarum* Cke. & Harkn. Pycnidia are black, globose, separate, erumpent, and ostiolate. Conidia are small, one-celled, and ovoid or ellipsoid. The olive brown conidia are approximately half the size of those of *C. wernsdorffiae* and are released in sooty masses beneath the epidermis of the rose canes. Pycnidia may even be found within black spot lesions on leaves, but common canker is primarily a disease of rose canes.

The pathogen rapidly colonizes wounded rose stems. Stem cankers are generally formed at the cut end of canes when stubs are left after pruning, but they may also form around insect punctures, thorn pricks, leaf or thorn scars, or abrasions caused by tying.

Control

Care should be taken to avoid injury to rose canes. Pruning cuts should be made immediately above the node without actually cutting the nodal tissue and at an angle to leave the minimum stub of dead wood. Rose stubs usually die back to the first node, and the pathogen invades and develops readily in such dead or dying tissue. Cuts made close to a node will quickly callus and prevent infection by *C. fuckelii*.

Infected canes should be removed at any node below the visibly diseased areas. Sharp pruning tools should be used to obtain clean cuts and avoid wounding. Fungicidal sprays may be used to cover wounds.

Selected References

Horst, R. K. 1979. *Coniothyrium*. Pages 157-158 in: Westcott's Plant Disease Handbook. 4th ed. Revised by R. K. Horst. Van Nostrand Reinhold Company, New York. 803 pp.

Matta, A., Garibaldi, A., and Gullins, G. 1976. Study of rose cane cankers in Piedmont and Liguria, Italy. Riv. Patol. Veg. 12:5-19.

Nelson, P. E., and Nichols, L. P. 1969. Root and stem diseases. Pages 203-210 in: Roses: A Manual on the Culture, Management, Diseases, Insects, Economics and Breeding of Greenhouse Roses. J. W. Mastalerz and R. W. Langhans, eds. Penn. Flower Growers, N.Y. State Flower Growers Assoc., Inc., and Roses Inc. 331 pp.

Brown Canker

Brown canker occurs primarily on outdoor-grown roses and may be widespread and serious wherever roses are grown. The disease was first reported in the eastern United States in 1917.

Symptoms

Initial symptoms of brown canker are small red to purple spots on the current year's canes. These spots enlarge into whitish necrotic lesions that scarcely penetrate the surface of the stem (Fig. 7). The lesions

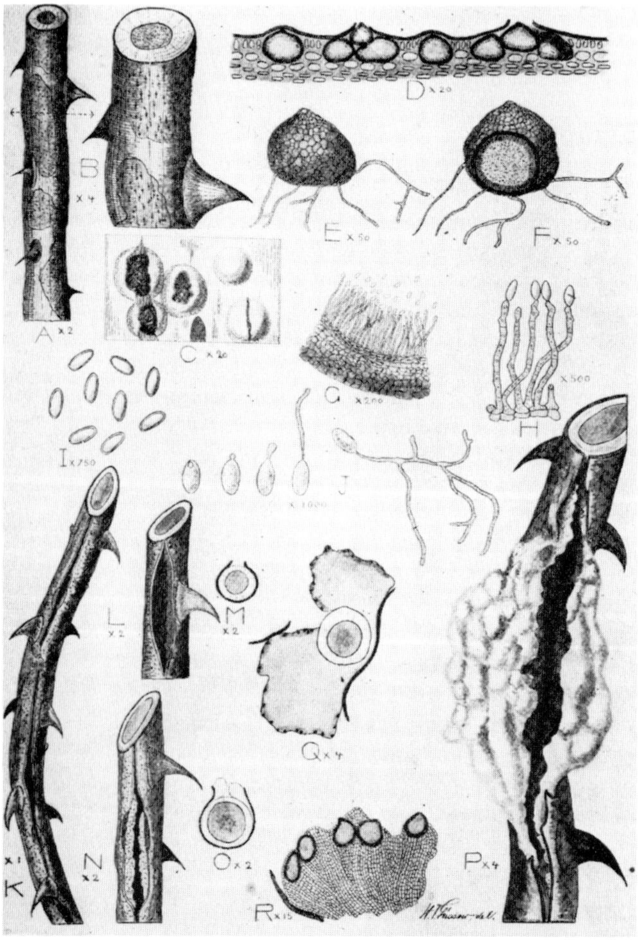

Fig. 6. Common canker caused by *Coniothyrium fuckelii*: cankers on cane (A); minute black pycnidia in cankered area (B); enlargement of pycnidia (C); longitudinal view of bark, showing slit that is formed as pycnidia increase in size (D); pycnidia with mycelia developing from base (E and F); conidiophores, on which conidia are borne within pycnidia (G and H); conidia (I); germinating conidia (J); slits in bark of canes with developing pycnidia (K, L, and N); cross sections through canes with slits in bark (M, O, and Q); galls that sometimes develop around cankered canes (P); cross section through bark, showing pycnidia (R). (Reprinted, by permission, from "Parasitic rose canker. A new disease in roses," by H. T. Gussow, 1908, J. R. Hortic. Soc. 34:222-230)

Fig. 7. Brown canker caused by *Cryptosporella umbrina*, showing light-colored canker on cane with minute black pycnidia in the cankered area. (Reprinted, by permission, from *Westcott's Plant Disease Handbook*, p. 160. 4th ed. Revised by R. K. Horst. 1979, Van Nostrand Reinhold Company, New York)

have a reddish purple margin. Cankers may be grouped together so that solid tan patches with purple borders are formed. During winter and spring, cankers that are several inches long develop on one-year-old canes. In moist weather, these large cankers are covered with yellow spore tendrils from pycnidia just beneath the epidermis; asci may also be extruded in tendrils from perithecia.

Causal Organism

Brown canker is caused by *Cryptosporella umbrina* (Jenkins) Jenkins & Wehm. The imperfect stage of this fungus, *Diaporthe umbrina* Jenkins, is often found in the cankers.

Control

Canes with cankers and dying stems should be pruned as soon as symptoms are observed. Spring pruning and flower cuts should be made immediately above buds or leaf axils. Fungicide sprays may be used to cover wounds.

Selected References

Horst, R. K. 1979. *Cryptosporella*. Pages 159-161 in: Westcott's Plant Disease Handbook. 4th ed. Revised by R. K. Horst. Van Nostrand Reinhold Company, New York. 803 pp.

Jenkins, A. E. 1927. Brown canker of the rose. Am. Rose Annu. 12:161-183.

Nelson, P. E., and Nichols, L. P. 1969. Root and stem diseases. Pages 203-210 in: Roses: A Manual on the Culture, Management, Diseases, Insects, Economics and Breeding of Greenhouse Roses. J. W. Mastalerz and R. W. Langhans, eds. Penn. Flower Growers, N.Y. State Flower Growers Assoc., Inc., and Roses Inc. 331 pp.

Black Mold

Black mold disease was first observed on roses in Australia in 1936 and was reported in the eastern United States in 1938 and 1939. The disease was reported in Scotland in 1980.

Symptoms

The heavy black mycelium of the causal fungus grows over and covers recently cut surfaces of stock and scion. This heavy growth prevents union of stock and scion, resulting in death of scions. Similar losses may occur on budded plants.

Newly infected rose grafts initially show a white to grayish white growth over the cut surface of stocks and scions. This mycelial growth gradually darkens until masses of black spores are produced; these give the disease its name. Some discoloration of scion and stock stems occurs, and callus formation is prevented.

On budded plants, initial symptoms are distinguished by the failure of buds to unite with stocks. Black mycelia and spores form on the interfaces of the bud shield and rootstocks. The earliest symptoms, which are somewhat inconspicuous and are frequently overlooked, consist of frosty white mycelial growth over the cut surfaces. High humidity is conducive to this fungus growth. The mycelium darkens in a few days, and masses of black spores are formed. The bud shield is ultimately killed.

The most severe symptoms occur on dormant or senescent wood such as bud shields, scions, cuttings, and branch stubs. Cuttings may be prevented from forming callus and rooting.

Causal Organism

Black mold is caused by the fungus *Chalaropsis thielavioides* Peyronel. The fungus is strictly a wound pathogen. Two types of conidia are produced. Macroconidia (Fig. 8B) or chlamydospores (Fig. 8A) are olive green and thick-walled when mature; these conidia are sessile or are borne on short conidiophores in compact groups. Hyaline endoconidia are formed inside end cells of dark endoconidiophores and are extruded in chains (Fig. 8C). Growth of this fungus on rose tissue in moist chambers produces a characteristic sweet, fruity odor.

Control

No control methods have been described for black mold. Species of rose used as understocks vary in susceptibility to *C. thielavioides*. *Rosa odorata* and *R. manetti* are very susceptible; *R. multiflora* is moderately susceptible. Among *R. multiflora* cultivars, Ragged Robin is immune, whereas Dr. Huey is susceptible.

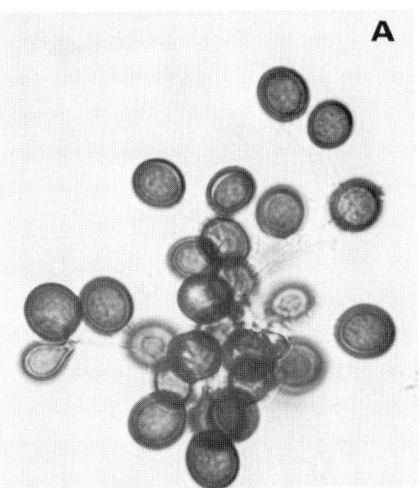
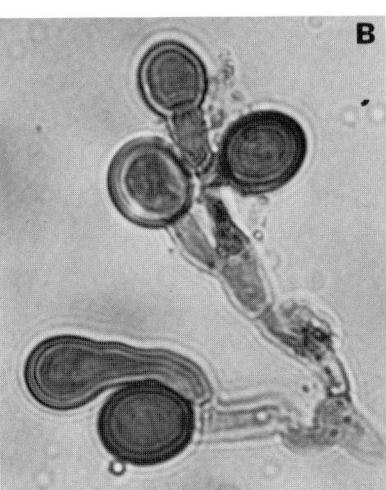
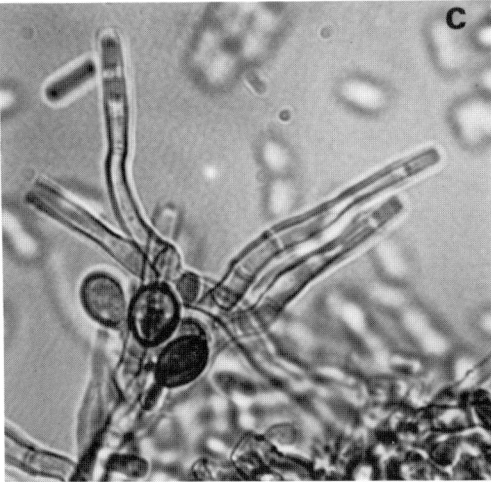

Fig. 8. Chlamydospores (**A**), conidia and conidiophores (**B**), and endoconidia (**C**) of *Chalaropsis thielavioides*, cause of black mold.

Selected References

Brokenshire, T. 1980. Black mold of roses in Scotland, UK caused by *Chalaropsis thielavioides*. Plant Pathol. 29:56.

Horst, R. K. 1979. *Chalaropsis*. Page 246 in: Westcott's Plant Disease Handbook. 4th ed. Revised by R. K. Horst. Van Nostrand Reinhold Company, New York. 803 pp.

Longrée, K. 1940. *Chalaropsis thielavioides*, cause of "black mold" of rose grafts. Phytopathology 30:793-807.

Massey, L. M., and Longrée, K. 1939. Black mold disease of rose grafts. Florists Exch. Hortic. Trade World 93:17.

Nelson, P. E., and Nichols, L. P. 1969. Root and stem diseases. Pages 203-210 in: Roses: A Manual on the Culture, Management, Diseases, Insects, Economics and Breeding of Greenhouse Roses. J. W. Mastalerz and R. W. Langhans, eds. Penn. Flower Growers, N.Y. State Flower Growers Assoc., Inc., and Roses Inc. 331 pp.

Botrytis Blight

Botrytis blight is widely distributed around the world on a great many flowers, fruits, and vegetables. The disease has numerous names, including gray mold blight, bud and flower blight, blossom blight, and gray mold rot. On roses, Botrytis diseases may also include cane canker. Botrytis blight occurs wherever outdoor and glasshouse roses are grown and has recently been reported to be troublesome in the United States, Iraq, Japan, India, and Canada.

Symptoms

The most severe damage from Botrytis blight occurs in storage or in transit. Infections may not be visible at the time of flower cut but develop rapidly in the humid conditions found in storage and during shipment. Procedures that maintain high moisture in shipping boxes provide ideal conditions for development of the pathogen.

During periods of continued wet weather and cool temperatures, buds of infected garden roses fail to open and become covered with the grayish brown mycelial growth of the causal fungus (Plate 30). Infected buds may droop, and smooth, slightly sunken, grayish black lesions may be found extending down the stem from the base of the bud.

Damage on glasshouse roses may be similar to that on outdoor roses or may appear as bruises. Small flecks appear on infected petals (Plates 31 and 32), and petal tips or sides become brown and soft (Plate 33). In some instances, numerous circular brown spots or blisterlike patches may appear over the surface of petals. Infections are especially obvious on cultivars with white flowers.

Botrytis may infect the stub ends from which flowers have been cut or the wounds from pruning in both glasshouse and field-grown roses. Ultimately, these infections result in blighting of canes (Fig. 9). Cankers may develop from Botrytis infections wherever wounds are made and when moisture and temperature conditions are conducive for development of the fungus. If conditions are favorable in the spring, new canes are often infected at nodes, with resultant girdling and collapse of stems. On cuttings in the propagation glasshouse, the fungus may gain entrance through the cutting wounds and kill small twigs or the entire cutting. Affected areas of plants are often covered with grayish brown mycelial growth and powdery masses of gray, airborne conidia (Plate 34).

Dormant rose plants held in storage are often infected with *Botrytis*. The entire plant becomes fuzzy with the growth of the fungus, and buds or large portions of plants are killed. Young canes are affected most frequently (Plate 35).

Causal Organism

Botrytis blight is caused by *Botrytis cinerea* Pers. ex Fr. This fungus undoubtedly has many strains, and perhaps more than one species infects roses, but they have not been thoroughly investigated and separated.

The optimal temperature for growth of the fungus and development of disease is 15°C, and high humidity and moisture are also required. The fungus usually requires a wound to invade tissue.

Conidia are egg-shaped, hyaline, and one-celled. They are formed on branched conidiophores (Fig. 10B) over the surface of infected tissue. There are no special fruiting bodies. The arrangement of conidia gives the

Fig. 9. Cane blight caused by *Botrytis cinerea*. Blight develops after the flower has been cut. Note dark sclerotia apparent in lesion.

genus its name; the Greek derivative *botrys* means "cluster of grapes" (Fig. 10A). Flattened, loaf-shaped, or hemispheric black sclerotia may be formed on or just beneath the cuticle or epidermis of the host and are firmly attached. These sclerotia, which have a dark rind and light interior, serve as resting bodies for overwintering.

Control

All infected buds, flowers, and canes in glasshouses, fields, and gardens should be cut and destroyed as soon as the first symptoms of Botrytis blight appear. Prompt removal prevents the formation of large numbers of *Botrytis* conidia that can be transported by air currents. Sanitation in rose-grading areas and storage coolers is very important.

Moisture condensation in glasshouses is greatest when outside temperatures are falling and the general humidity is high. Condensation can be reduced by ventilation and air circulation. Good ventilation should be provided for roses in storage and in propagating beds.

Protective fungicidal sprays should be used to cover wounds. Stored roses should be sprayed or dipped when they are brought into storage. A major problem associated with fungicidal treatments is the development of resistance in *B. cinerea*. Fungicides are often ineffective against *B. cinerea* after only three seasons of spraying or even as early as three spray applications.

Botrytis sporulates best at light wavelengths of 355 nm (ultraviolet light). Some types of greenhouse covers will filter this wavelength of light. Thus, ultraviolet-absorbing vinyl film has been suggested to give effective control of *Botrytis* sporulation.

Selected References

Chohan, J. S., and Kaur, S. 1976. Gray mold and *Pestolotiosis laprogena* rot on rose buds and flowers. Indian Phytopathol. 29:98.

Damirdagh, I. S. 1979. Botrytis blight of roses in Sulaimaniya, Iraq. (Abstr.) Phytopathology 69:1026.

Gupta, G. K. 1979. Bud and twig blight of rose caused by *Botrytis cinerea* Pers. ex Fr. Indian J. Hortic. 36:473.

Hisada, Y., Takaki, H., Kawase, Y., and Ozaki, T. 1979. Difference in the potential of *Botrytis cinerea* to develop resistance to procymidone in vitro and in the field. Ann. Phytopathol. Soc. Jpn. 45:283-290.

Honda, Y., Toki, T., and Yunoki, T. 1977. Control of gray mold of greenhouse cucumber and tomato by inhibiting sporulation. Plant Dis. Rep. 61:1041-1044.

Horst, R. K. 1979. *Botrytis*. Pages 110-112 in: Westcott's Plant Disease Handbook. 4th ed. Revised by R. K. Horst. Van Nostrand Reinhold Company, New York. 803 pp.

Jarvis, W. R., and Slingsby, K. 1975. Tolerance of *Botrytis cinerea* and rose powdery mildew to benomyl. Can. Plant Dis. Surv. 55:44.

Nichols, L. P., and Nelson, P. E. 1969. Foliage diseases. Pages 185-195 in: Roses: A Manual on the Culture, Management, Diseases, Insects, Economics and Breeding of Greenhouse Roses. J. W. Mastalerz and R. W. Langhans, eds. Penn. Flower Growers, N.Y. State Flower Growers Assoc., Inc., and Roses Inc. 331 pp.

Canker (Dieback)

Several fungi besides those described in the preceding sections have been reported to cause canker or dieback disease of rose. The importance of these pathogens has not been well documented, nor has the occurrence of these diseases been thoroughly investigated. The individual pathogens and the symptoms that have been reported in connection with them are described in this section.

Botryosphaeria ribis Gross & Dug. is a saprophyte that develops on dying tissue; the parasitic form *B. ribis* var. *chromogena* Shear, N. E. Stevens, & M. S. Wilcox causes symptoms late in the season that include small, dark, wartlike fruiting bodies in definite parallel rows on diseased canes. Rose canes wilt and die back above cankers. *B. ribis* var. *chromogena* develops a purple-pink color when grown on starch paste.

Cryptosporium minimum Laub. has been reported to cause cankers on rose but is not commonly found.

Griphosphaeria corticola (Fckl.) Hoehn. (imperfect stage *Coryneopsis microsticta* (Berk. & Br.) Grove) causes cankers at the base of canes (Fig. 11). Dark, glistening pustules of conidia may be seen. When canes are girdled, large galls may be formed above cankers (Fig. 11). These

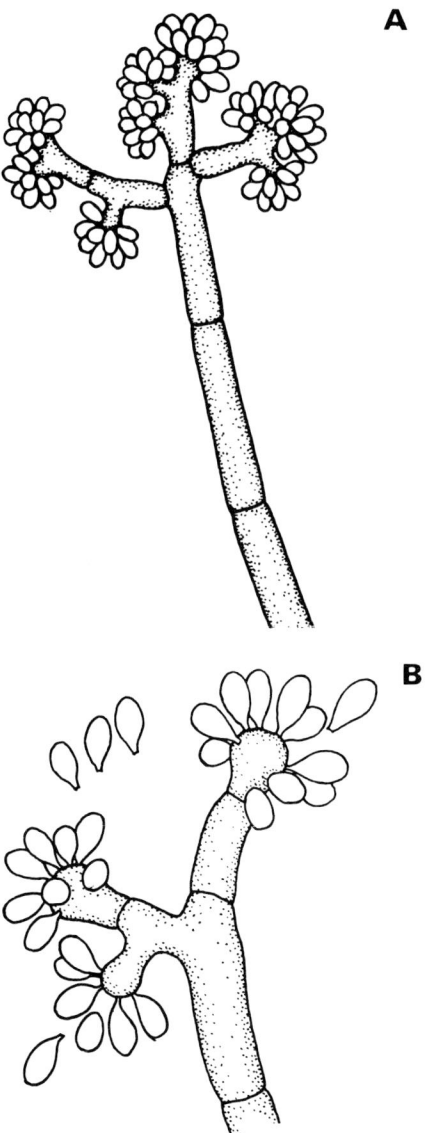

Fig. 10. **A,** Branched conidiophores of *Botrytis cinerea* resemble bunches of grapes. **B,** Conidia of *B. cinerea* and their attachment to conidiophores.

galls resemble crown gall and are apparently caused by interference with the downward transport of food.

Nectria cinnabarina Tode ex Fr. is widespread in occurrence as a saprophyte and is weakly parasitic on rose. Cankers are produced around wounds and at the base of dead branches (Plates 36 and 37).

The symptoms associated with *Didymella sepincoliformis* (de N.) Sacc. are described as dieback. The symptoms associated with *Glomerella cingulata* Penz. are described as dieback and canker.

Cylindrocladium scoparium Morg. causes symptoms below the union of stock and scion. The bark darkens into a black, water-soaked, punky region (Fig. 12). Cankers girdle but do not kill canes; however, fewer and inferior flowers are produced from these canes.

Disease caused by *Diaporthe eres* Nits. (imperfect stage *Phomopsis mali* (Schultz & Sacc.) Roberts) has been reported in the United States and Italy and is particularly serious on bush roses and tree roses grown on the understocks of *Rosa canina* and *R. multiflora*. Dark cankers found on canes may contain pycnidia of the pathogen.

Selected Reference

Horst, R. K. 1979. Cankers and diebacks, and Rose. Pages 153-181 and 669-671 in: Westcott's Plant Disease Handbook. 4th ed. Revised by R. K. Horst. Van Nostrand Reinhold Company, New York. 803 pp.

Miscellaneous Diseases Caused by Fungi

Several rose diseases caused by fungi have been reported for which little information is available. The occurrence of these diseases has not been thoroughly investigated, nor has the importance of the causal fungi been well documented.

Spot Anthracnose

Sphaceloma rosarum (Pass.) Jenkins (perfect state *Elsinoë rosarum*), the cause of spot anthracnose, was collected from wild roses as early as 1898. It was reported as more serious on climbing and rambler roses but was also pathogenic on hybrid tea and bush roses. The disease has been confused with black spot and when conditions are favorable can be just as serious, causing spotting, yellowing, and defoliation.

Leaf spots are scattered or grouped, sometimes running together; they are usually circular and up to 0.5 cm in diameter (Fig. 13A). Young spots are red, varying to brown or dark purple on upper leaf surfaces. Ultimately, the centers of spots turn ashen white with a dark red margin (Fig. 13B). Tiny acervuli of the fungus

Fig. 11. Canker and gall caused by *Griphosphaeria corticola*. (Reprinted, by permission, from *Westcott's Plant Disease Handbook*, p. 167. 4th ed. Revised by R. K. Horst. 1979, Van Nostrand Reinhold Company, New York)

Fig. 12. Cane canker **(A)** and cracking of bark in cankered areas **(B)** caused by *Cylindrocladium scoparium*.

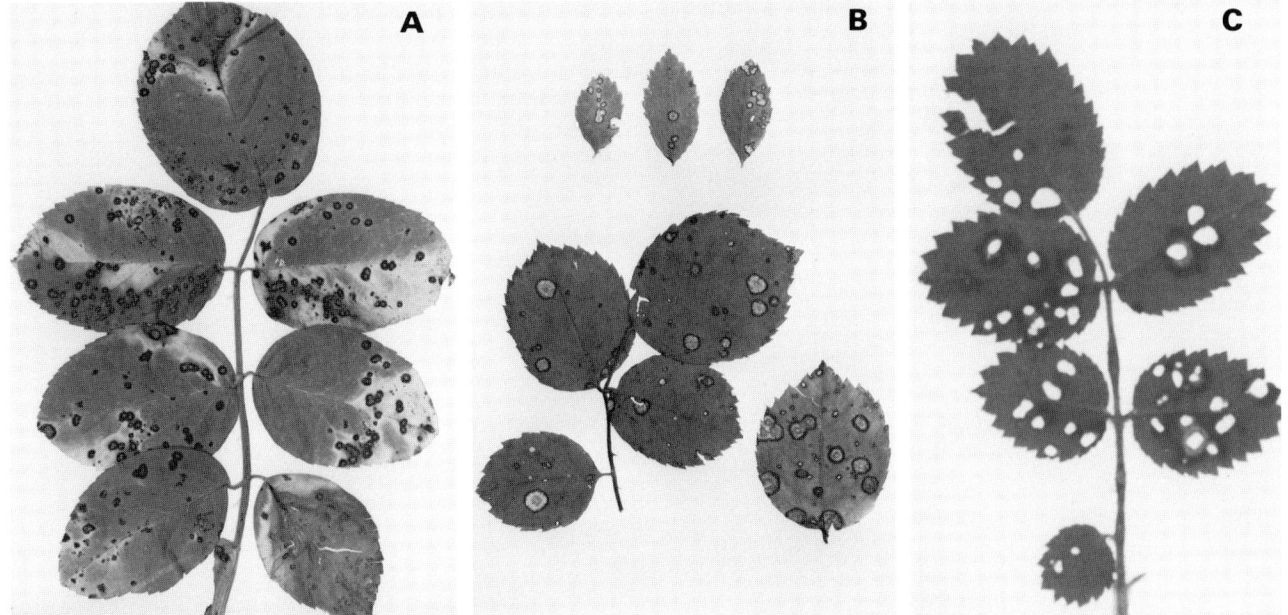

Fig. 13. Spot anthracnose caused by *Sphaceloma rosarum*. **A,** Grouped leaf spots on leaf surface. **B,** Light-colored leaf spots with dark margins. **C,** Shot-hole symptom where tissue in centers of spots has fallen away.

may appear scattered on the white centers of spots. Tissue in the lesions may fall away from the lower leaf surface, leaving a thin, papery membrane, or it may fall out completely, producing a shot-hole symptom (Fig. 13C). Spots with ashen centers may also be found on stems, rose hips, and pedicels.

Conidia of *S. rosarum* are formed in early spring and continue to form through summer when wet conditions prevail. Conidia are spread by splashing water.

Wilt

Phytophthora megasperma Drechsler causes wilt in poorly drained areas. Wounds are necessary for infection. Stems near the soil surface become water-soaked, dark green, and finally dark brown. New shoots wilt and die. Lower leaves of older plants yellow, wilt, and finally abscise. Cultivars known to be particularly susceptible are Carina, Golden Rapture, and Mary Devor. Plants grown from cuttings are more susceptible than those grafted to *Rosa multiflora*.

Leaf Spots

Alternaria alternata (Fr.) Keissler causes leaf spot symptoms during rainy periods. After initial spotting, leaves become brittle and change from yellow-brown to dark brown. Spots enlarge and show concentric rings on ridges. If moist, humid conditions persist, flower buds and flowers may become infected. The optimal temperature for disease is 30°C.

A. brassicae var. *microspora* Brun. and other species of *Alternaria* have also been reported to cause leaf spots on rose.

Cercospora puderi B. H. Davis causes circular spots up to 5 mm in diameter with dull gray centers and brown to reddish brown margins. The fungus sporulates mainly on upper leaf surfaces in dense fascicles of short conidia.

C. rosicola Pass. (perfect state *Mycosphaerella rosicola* B. H. Davis) causes circular spots 1–4 mm in diameter that coalesce to form irregular, purple to

Fig. 14. Cercospora leaf spot caused by *Cercospora puderi* and *C. rosicola*. The circular spots sometimes coalesce.

reddish brown areas (Fig. 14) with pale brown, tan, or gray centers (Plate 38). Perithecia are formed in fallen leaves. *M. rosigena* (Ell. & Ev.) Lindau has also been reported as a rose pathogen but has probably been confused with *M. rosicola*. Conidia of *C. puderi* and *C. rosicola* are compared in Figure 15.

Colletotrichum capsici (Syd.) Butl. & Bisby causes circular, oxblood red spots. Initially small, the spots gradually enlarge and turn chestnut brown. Spots may coalesce until major portions of leaves are affected. Leaves finally dry out and fall.

Other fungi that have been isolated from rose leaf spots include *Monochaetia compta* (Sacc.) Allesch., *Pezizella oenotherae* (Cke. & Ell.) Sacc., *Phyllosticta rosae* Desm., *Septoria rosae* Desm., *Glomerella cingulata* Penz., and *Curvularia brachyspora* Boedijn.

Petal Spots

Bipolaris (Helminthosporium) setariae (Saw.) Shoemaker initially causes tan spots up to 2 mm in diameter. Spots gradually expand and coalesce, resulting in large, tan, necrotic areas on petals. Severely infested petals fall from flowers. High moisture and humidity are required for disease to develop. Symptoms resemble those caused by *Botrytis*.

Blight

Physalospora fusca N. E. Stevens, *P. obtusa* (Schw.) Cke., *Gloeosporium rosaecola* Dearn. & Barth. or *G. rosarum* (Pass.) Grove (imperfect state *Sphaceloma*), *Pellicularia koleroga* Cke., and *Sclerotium rolfsii* Sacc. have been reported to cause blights of rose.

Root Rots

Armillaria mellea Vahl ex Fr., *Clitocybe tabescens* (Scop. ex Fr.) Bres., *Phymatotrichum omnivorum* (Shear) Dug., and *Ramularia macrospora* Fres. have been reported to cause root rots of rose.

Algal Leaf and Stem Spot

The algal pathogen *Cephaleuros virescens* Kunge has been reported on 145 plant species in 51 families. Most leaf infections are economically insignificant. However, stem infections have been reported on rose. Their economic significance has not been established.

Selected References

Bedi, P. S., and Singh, J. P. 1972. Leaf blight of rose in the Punjab. Indian Phytopathol. 25:534-539.

Engelhard, A. W. 1976. Pathogenicity and conditions for infection of chrysanthemum and rose flowers by *Bipolaris (Helminthosporium) setariae*. Phytopathology 66:389-391.

Holcomb, G. E. 1976. Economically important hosts susceptible to stem infections by the alga *Cephaleuros virescens*. (Abstr.) Proc. Am. Phytopathol. Soc. 3:337.

Horst, R. K. 1979. Rose (*Rosa*). Pages 669-671 in: Westcott's Plant Disease Handbook. 4th ed. Revised by R. K. Horst. Van Nostrand Reinhold Company, New York. 803 pp.

Khanna, K. K., and Chandra, S. 1977. Control of leaf blight of *Rosa indica* and *Cinnamomum camphora* caused by *Glomerella cingulata*. Indian J. Mycol. Plant Pathol. 7:176-177.

Kore, S. S., and Bhide, V. P. 1976. A first report of *Curvularia brachyspora* Boedijn. inciting leaf-spot disease of rose. Curr. Sci. 45:74.

Nagai, Y., Takeuchi, T., and Watanabe, T. 1978. A stem blight of rose caused by *Phytophthora megasperma*. Phytopathology 68:684-688.

Nichols, L. P., and Nelson, P. E. 1969. Foliage diseases. Pages 185-195 in: Roses: A Manual on the Culture, Management, Diseases, Insects, Economics and Breeding of Greenhouse

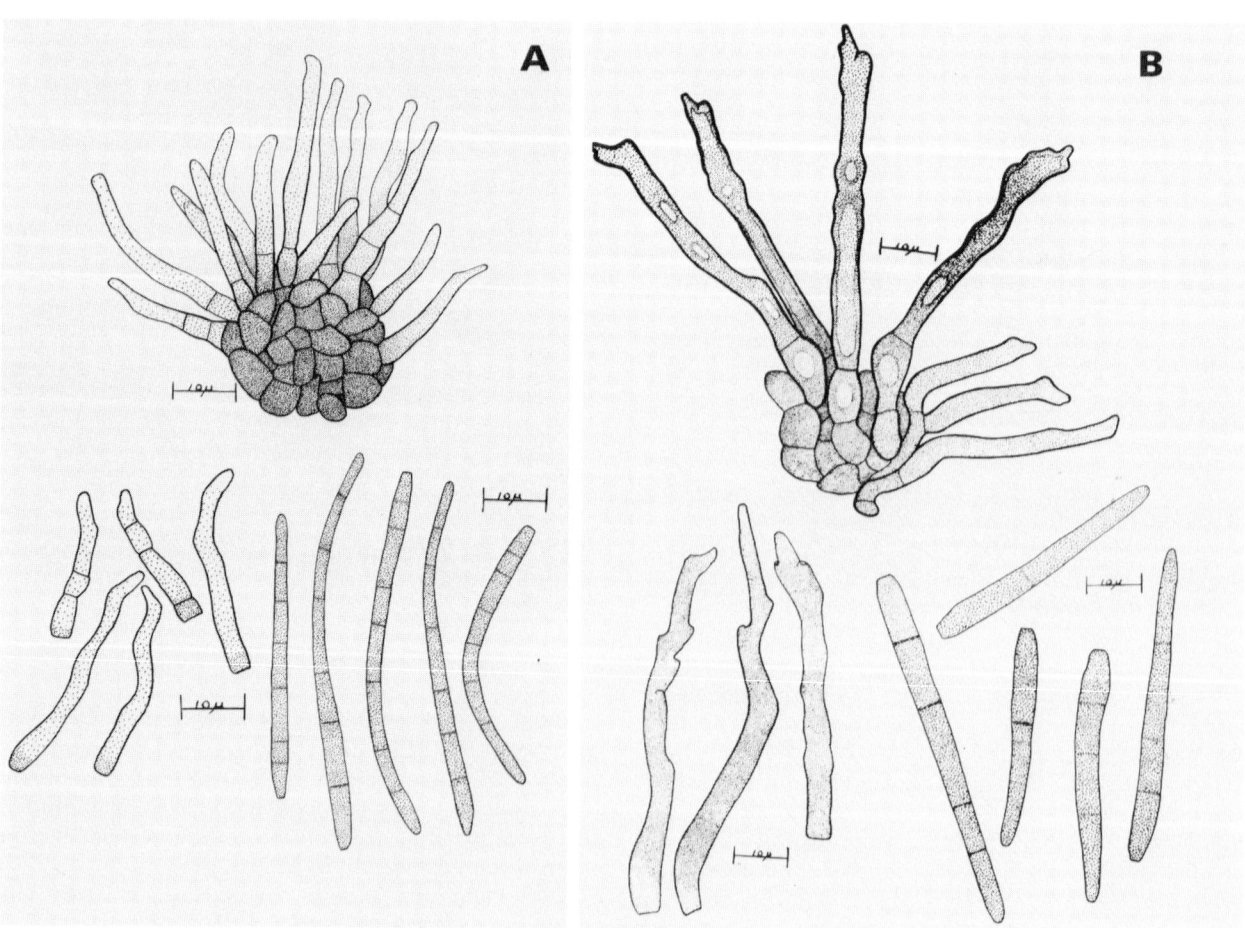

Fig. 15. Conidia and conidiophores of *Cercospora puderi* (A) and *C. rosicola* (B).

1–7. Powdery mildew caused by *Sphaerotheca pannosa*. **1,** Leaf curl and distortion symptoms. **2,** Heavy white mycelial growth of the fungus on upper leaf surfaces. **3,** Symptoms on cane and thorns. **4,** Mycelial growth of the fungus on calyx of flower bud. **5,** Germinating conidia of the fungus. **6,** Mycelial growth of the fungus on leaf surface. **7,** Conidia of the fungus forming on the ends of conidiophores. (Plate 1 reprinted, by permission, from *Westcott's Plant Disease Handbook*. 4th ed. Revised by R. K. Horst. 1979, Van Nostrand Reinhold Company, New York; Plates 5–7 courtesy of D. L. Coyier)

8-11. Black spot caused by *Diplocarpon rosae*. **8,** Early lesions on leaf. **9,** Acervuli of the fungus in lesion with feathery margins. **10,** Close-up of acervuli in lesion on leaf. **11,** Severe symptoms of black spot. Note lesions with yellowing from formation of ethylene.

12–19. Rust caused by *Phragmidium mucronatum*. **12,** Orange pustules of uredial stage on lower leaf surface. **13,** Close-up of uredial stage on lower leaf surface. **14,** Chlorotic spots on upper leaf surface. **15,** Uredial stage on cane. **16,** Black pustules of telial stage and orange pustules of uredial stage on leaf surfaces. **17,** Teliospores of the fungus. **18,** Telial stage and uredial stage on the same leaf. **19,** Overwintering telial stage on canes. (Plates 12 and 13 reprinted, by permission, from *Westcott's Plant Disease Handbook*. 4th ed. Revised by R. K. Horst. 1979, Van Nostrand Reinhold Company, New York; Plate 14 courtesy of K. Ohkawa; Plate 16 courtesy of V. A. Wager; Plates 17 and 18 courtesy of K. Milne)

20–25. Verticillium wilt caused by *Verticillium* spp. **20,** General wilting and yellowing of leaves. **21,** General wilting of cane with flower bud. **22,** Chlorotic leaves after infection. **23,** Yellowed canes and dark streaks after infection. **24,** Close-up of yellowed canes with dark streaks. **25,** Conidia of *Verticillium* in xylem of infected canes. (Plates 20 and 25 courtesy of C. Harwood)

26–29. Downy mildew caused by *Peronospora sparsa*. **26,** Purplish red to dark brown irregular spots on upper leaf surface. **27,** Close-up of spots on upper leaf surface. **28,** Necrotic lesions on upper leaf surface. **29,** Mycelia and conidia on lower leaf surface beneath lesion.

30-35. Botrytis blight caused by *Botrytis cinerea*. **30,** Unopened flower bud with fuzzy mycelial growth and sporulation of the fungus. **31,** Small flecks on flower petals. **32,** Petal lesions. **33,** Brown soft rot of flower petals. **34,** Cane blight with fuzzy mycelial growth and sporulation of the fungus. **35,** Blight of terminal growing region of cane. (Plates 31 and 32 courtesy of K. Welch, Mallinckrodt, Inc., Oakland, CA; Plate 33 reprinted, by permission, from *Westcott's Plant Disease Handbook*. 4th ed. Revised by R. K. Horst. 1979, Van Nostrand Reinhold Company, New York; Plate 35 courtesy of D. L. Coyier)

36 and 37. Canker caused by *Nectria cinnabarina*. **36,** Cane canker with sporodochia of the fungus. **37,** Close-up of sporodochia of the fungus.

38. Cercospora leaf spot. Circular spots have purplish to reddish brown margins and grayish centers. (Courtesy V. A. Wager)

39–42. Crown gall caused by *Agrobacterium tumefaciens*. **39,** Gall at crown of plant or just below soil surface. **40 and 41,** Galls on aerial portion of cane. **42,** Cut gall, showing lighter internal tissue and darkened, woody, outer portions of gall that slough off. (Plate 39 courtesy of L. W. Moore)

43–50. Rose mosaic. **43,** Foliar symptoms, including line patterns, ring spots, and mottles. **44,** Line pattern symptom on leaf. **45** and **46,** Ring spot symptom on leaves. **47,** Mottle symptom on leaf. **48,** Yellow net symptom on leaf. **49,** Yellow mosaic symptom on leaf. **50,** Portion of a plant with yellow mosaic symptoms. (Plates 43, 45, 46, 49, and 50 reprinted, by permission, from "Rose virus and virus-like diseases," by G. A. Secor, M. Kong, and G. Nyland, 1977, Calif. Agric. 31(3):4-7; Plate 44 reprinted, by permission, from *Westcott's Plant Disease Handbook*. 4th ed. Revised by R. K. Horst. 1979, Van Nostrand Reinhold Company, New York)

51–54. Rose ring pattern. **51,** Foliar distortion on *Rosa multiflora*. **52,** Foliar distortion (healthy leaf at right). **53,** Line pattern on leaf. **54,** Color break in petals. (Reprinted, by permission, from "Rose virus and virus-like diseases," by G. A. Secor, M. Kong, and G. Nyland, 1977, Calif. Agric. 31(3):4-7)

55-58. Rose spring dwarf. **55,** Leaf rosette following bud break. **56,** Close-up of leaf rosette. **57,** Zigzag manner of cane growth. **58,** Leaf balling symptoms on *Rosa multiflora* 'Burr'. (Reprinted, by permission, from "Rose virus and virus-like diseases," by G. A. Secor, M. Kong, and G. Nyland, 1977, Calif. Agric. 31(3):4-7)

59-62. Rose leaf curl. **59,** Leaf epinasty and shoot necrosis. **60,** Leaf epinasty. **61,** Vein flecking. **62,** Stem pitting in cane. (Reprinted, by permission, from "Rose virus and virus-like diseases," by G. A. Secor, M. Kong, and G. Nyland, 1977, Calif. Agric. 31(3):4-7; Plate 61 was published in black and white in "Rose leaf curl, a distinct component of a disease complex which resembles rose wilt," by S. A. Slack, J. A. Traylor, H. E. Williams, and G. Nyland, 1976, Plant Dis. Rep. 60:178-182)

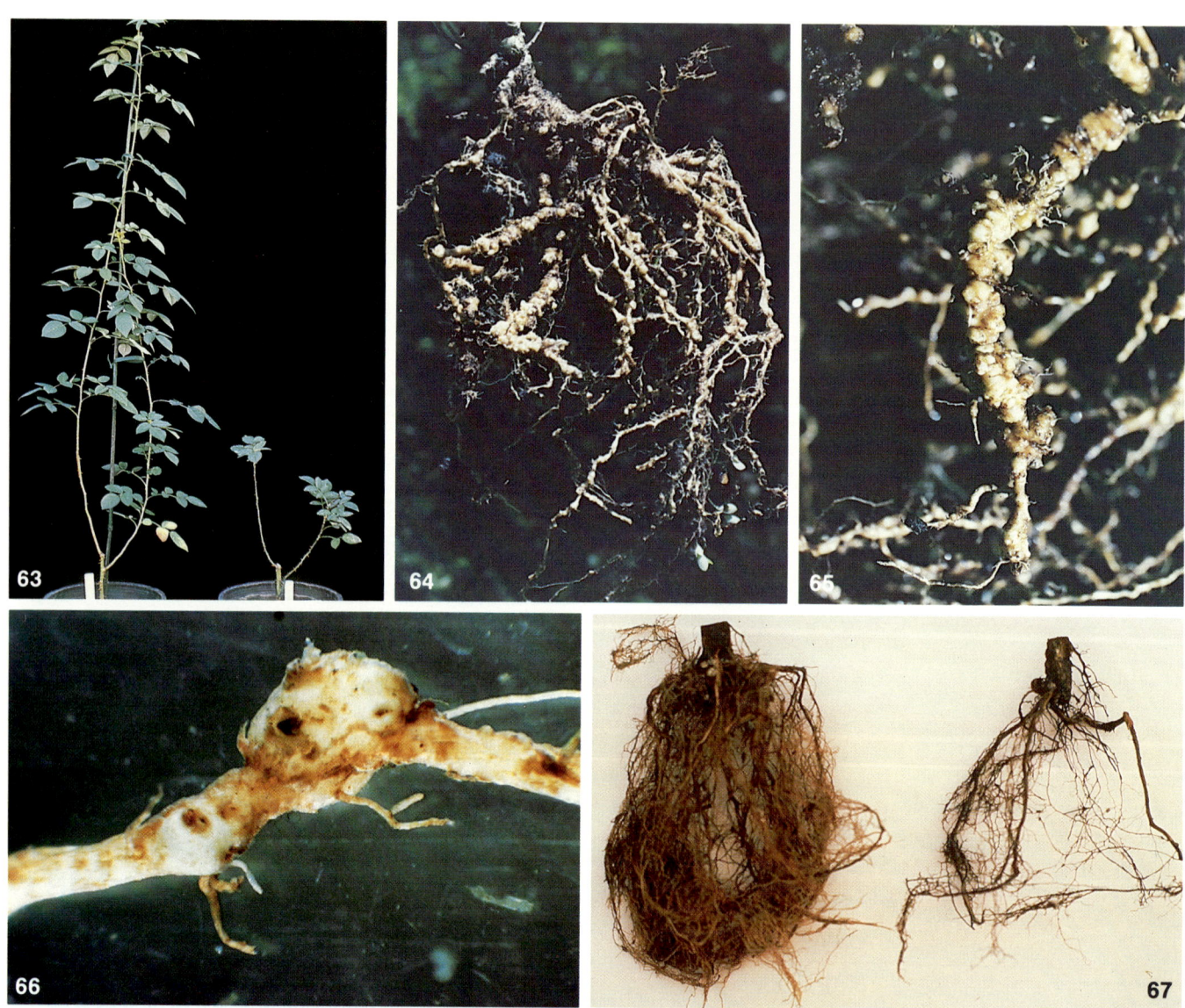

63–67. Disease symptoms caused by nematode infestations. **63,** General decline and stunting of plant (right), compared with a healthy plant (left). **64** and **65,** Galls on roots of gardenia infested with species of *Meloidogyne*. Symptoms resemble those found on rose. **66,** Close-up of root galls caused by infestation with *Meloidogyne*. **67,** Poor root growth of rose infested by nematodes (right), compared with healthy roots (left). (Plates 63 and 66 reprinted, by permission, from "Influence of *Pratylenchus vulnus* and *Meloidogyne hapla* on the growth of rootstock of rose," by G. S. Santo and B. Lear, 1976, J. Nematol. 8:18-23)

68. Blindness (failure of flower bud to form and/or abortion of terminal bud). (Courtesy M. Rogers)

69. Fluoride toxicity. Note marginal leaf chlorosis, thin cane, and dwarfed plant. (Courtesy R. Brewer, University of California, San Joaquin)

70. Ethylene damage. Note typical epinasty and curling of young leaflets. (Courtesy O. W. Davidson, Rutgers University)

71. Trifluralin damage. Note very small leaves formed after exposure. (Courtesy J. G. Seeley)

COLOR PLATES

72–83. Nutrient imbalance symptoms. **72,** Oxygen deficiency. Note yellowing of main veins and interveinal chlorosis. **73,** Nitrogen deficiency. Note overall yellow-green color of leaves. **74,** Phosphorus deficiency. Leaves are generally stunted and may be gray-green. **75,** Potassium deficiency. Note tip and marginal yellowing, browning, and necrosis of older leaves. **76,** Calcium deficiency. Leaves become dull gray-green and bend down at margins; edges later turn yellow and brown. **77,** Magnesium deficiency. Chlorosis of interveinal areas progresses to necrosis. **78,** Iron deficiency. Note interveinal chlorosis in which the main veins remain green. **79,** Zinc deficiency. Distorted leaves show yellowing that becomes necrotic. **80,** Manganese deficiency. Interveinal chlorosis becomes necrotic. **81,** Manganese toxicity. Note small black spots in interveinal areas on older leaves. **82,** Boron toxicity. Note marginal browning and necrosis of older leaves. **83,** Heat or salt stress. Note marginal necrosis of leaves. (Plates 72–78 and 80–83 reprinted, by permission, from "Greenhouse roses: Diagnosis and remedy of nutritional disorders," by J. W. White. 1976, Roses Inc., Haslett, MI; Plate 79 courtesy of C. Harwood)

Roses. J. W. Mastalerz and R. W. Langhans, eds. Penn. Flower Growers, N.Y. State Flower Growers Assoc., Inc., and Roses Inc. 331 pp.

Raizada, M., Khanna, K. K., and Chandra, S. 1975. A new leaf-spot disease of *Rosa indica* caused by *Colletotrichum capsici* and its control. Indian Phytopathol. 28:542.

Sahni, M. L. 1973. *Alternaria alternata* leaf blight of roses and its control through fungicidal sprays. Indian J. Mycol. Plant Pathol. 3:150-152.

Singh, J. P., and Bedi, P. S. 1980. Histopathological studies of leaf blight of rose caused by *Alternaria alternata*. Haryana Agric. Univ. J. Res. 10:333-335.

Diseases Caused by Bacteria

Crown Gall

Crown gall has been reported in all parts of the world on various plants. It was first observed on grape in Europe in 1853, and the bacterium was first isolated from galls on Paris daisy in the United States in 1904. Crown gall disease occurs on more than 60 dicotyledonous plant families, including Rosaceae. Some common hosts are almond, apple, apricot, aster, beet, blackberry, cherry, chrysanthemum, daisy, euonymus, fig, grape, nectarine, peach, pear, plum, raspberry, rose, walnut, and willow. The disease is found on pecan, shade trees, and other hardwoods; it is rarely found on conifers. Tomato, sunflower, and bryophyllum are widely used as test plants in research on crown gall. The disease occurs worldwide on roses and is commonly observed on roses in the United States.

Symptoms

Crown gall is primarily a disease of the parenchyma tissue, starting with a rapid proliferation of cells in meristematic tissues and the formation of more or less convoluted, soft or hard overgrowths or tumors. The overgrowths or galls are often found at or just below the soil surface in the basal or crown region (Fig. 16A and Plate 39) of plants (hence the name crown gall). They may occur frequently on roots (Fig. 16B) and less frequently on aerial plant parts (Plates 40 and 41).

The galls are usually rounded, with a rough, irregular surface. They first appear as small protuberances on the plant surface. Galls may vary from 0.5 cm to several centimeters in diameter, depending somewhat on the plant's vigor and growth and on the time after infection. Young, actively developing galls are light green or nearly white, and the tissue is soft. As they age, galls become darkened and woody (Plate 42). Outer portions of galls can slough off with age due to normal rotting, weathering, or the action of other microorganisms. Sometimes the galls have a rather smooth surface, which makes it difficult to distinguish between gall and callus growth, especially if the gall occurs at the base of plants or at the graft or bud union. Symptoms of crown gall differ from those of hairy root in that hairy root galls do not become woody, and small roots develop from hairy root galls and later form large masses of fibrous roots (see Hairy Root).

The actual economic importance of crown gall in rose production is difficult to assess. Losses due to stunting, poor foliage, and fewer blossoms caused by crown gall disease are difficult to separate from losses due to similar plant responses induced by other diseases or

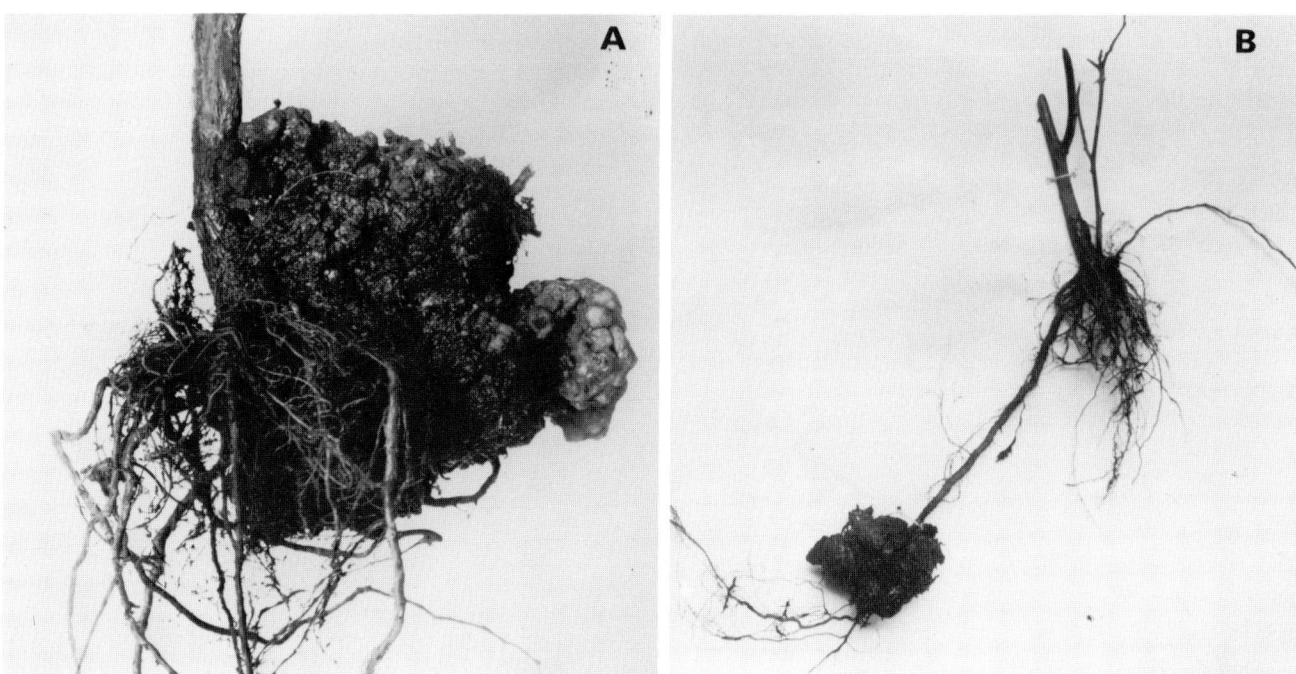

Fig. 16. Crown gall, caused by *Agrobacterium tumefaciens*, at crown or basal region of plant **(A)** and on roots **(B)**.

factors. Sometimes, plants severely infected by the crown gall bacterium show few adverse effects, while in other instances, plants with few galls may be a total loss. It is likely that the location of galls determines the effect of the disease on plants; a single gall at the base of a plant may be more detrimental than several galls on canes and roots.

Causal Organism

Crown gall is caused by the bacterial pathogen *Agrobacterium tumefaciens* (E. F. Sm. & Town.) Conn. This pathogen has been given several names in the past, including *Bacterium tumefaciens, Pseudomonas tumefaciens, Bacillus tumefaciens*, and *Phytomonas tumefaciens*. The bacterium is rod-shaped, 0.7–0.8 μm wide and 2.5–3.0 μm long, motile with one to four flagella, and gram-negative. It does not form spores.

Etiology and Epidemiology

Bacterial activity is greatest during the summer months. The pathogen enters plants through wounds, either natural or caused by pruning, grafts, mechanical injury from cultivation, "heaving" of frozen soils, chewing insects, or the emergence of lateral roots. Galls become visible a week to several months after infection.

After *A. tumefaciens* enters a wound, plasmid DNA transferred from the bacterium into the nuclear genome of the plant cells transforms normal cells within the wounded area into tumor cells. Once the cells are transformed, the abnormal proliferation of tumor cells becomes an autocatalytic process that continues independently of the bacterium. Anatomically, cells immediately surrounding the transformed cells usually divide rapidly. New growth centers may also form and contribute to the distorted structure of the gall. The orientation and size of proliferating cells vary, and vascular elements may differentiate as a continuation of existing xylem elements or as discontinuous groups. The end result is a gall composed of a more or less disorganized mass of hyperplastic and hypertrophic tissue.

The rate of gall development is influenced by type of host, vigor and growth of host, and environmental conditions. Infections in autumn months may remain latent, and dormant infected plants remaining in the field or in cold storage do not show symptoms until late the following spring.

Bacteria are generally more abundant between cells located at or near the surface of developing galls. Thus, pruning tools that cut through galls can become contaminated with the bacteria, spread them to cut surfaces of subsequently pruned plants, and thereby induce new infection.

As galls in soil break down, bacteria are released and can be transported by moving soil or water. In the absence of plant roots, bacterial populations gradually decrease; however, the pathogen may survive in soil for at least two years.

Control

Rootstocks differ in susceptibility to *A. tumefaciens*. *Rosa multiflora, R. manetti,* and Bayse No. 3 are very susceptible, whereas Iowa State University (ISU) 60-5, Brooks 48, Clarke 1957, and Welch are resistant. However, no rootstocks are known to be immune.

Several generally recommended practices reduce the occurrence of crown gall. First, use only disease-free plants; however, plants with latent infections appear to be disease-free even though the pathogen may be present as a resident. Second, avoid injury to roots and crown during planting and cultivating. Third, plant in soil that has been properly treated or sterilized. In some instances, increased crown gall on roses has been observed following fumigation with an ineffective chemical. Fourth, remove infected plants as soon as galls are observed. If possible, remove soil from around the area of infected plant roots to be sure that all galls and infected roots are discarded. Fifth, wash cutting and pruning tools thoroughly with soap and water and disinfest them frequently. To disinfest tools, dip them in denatured alcohol and flame them, or dip them for several minutes in 0.5% sodium hypochlorite solution and wash them in boiled water to prevent corrosion. And sixth, rotate crops, using a monocotyledonous cover crop.

Biological control of crown gall by the antagonistic bacterium *A. radiobacter* strain K84 has aroused interest in recent years. *A. radiobacter* strain K84 prevents sensitive strains of *A. tumefaciens* from initiating crown gall when the population of the antagonist in wound sites equals or preferably exceeds that of the pathogen. The antagonist must be used as a preventive measure rather than a method of eradicating the pathogen. Field experiments in Canada, France, Greece, Hungary, Italy, New Zealand, South Africa, and the United States have confirmed the effectiveness of *A. radiobacter* in controlling crown gall on a wide variety of plants.

Two possibilities must be kept in mind concerning the use of this biological control procedure: 1) the pathogenic bacterium might conjugate with *A. radiobacter* strain K84 and receive the agrocin 84 plasmid, which results in resistance to K84, so that it would no longer be subject to biological control by K84; and conversely, 2) strain K84 might receive the Ti plasmid (responsible for pathogenicity) from the pathogen donor, become pathogenic, and cause crown gall. There is evidence for the occurrence of both types of genetic transfer. However, in only one reported instance has such a genetic transfer resulted in a breakdown of biological control by K84. Strain K84 has been used successfully with roses in Australia, New Zealand, and Spain but has not been effective in the United States in the limited trials in which it has been tested. It is possible that *A. tumefaciens* strains immune to K84 may already exist in some fields.

An oil-water emulsion containing 2,4-xylenol and metacresol (Bacticin) has been eradicative when painted directly on galls. However, this type of treatment is slow and time-consuming and has limited application and effect, in part because tumor tissue often resumes growth after the treatment.

Selected References

Cooksey, D. A., and Moore, L. W. 1982. High frequency spontaneous mutations to agrocin 84 resistance in *Agrobacterium tumefaciens* and *A. rhizogenes*. Physiol. Plant Pathol. 20:129-135.

Dickey, R. S. 1969. Crown gall disease. Pages 196-202 in: Roses: A Manual on the Culture, Management, Diseases, Insects, Economics and Breeding of Greenhouse Roses. J. W. Mastalerz and R. W. Langhans, eds. Penn. Flower Growers,

N.Y. State Flower Growers Assoc., Inc., and Roses Inc. 331 pp.

Horst, R. K. 1979. *Agrobacterium tumefaciens*. Pages 74-76 in: Westcott's Plant Disease Handbook. 4th ed. Revised by R. K. Horst. Van Nostrand Reinhold Company, New York. 803 pp.

Kerr, A. 1980. Biological control of crown gall through production of agrocin 84. Plant Dis. 64:25-30.

Massey, L. M. 1950. Crown gall. Am. Rose Annu. 35:145-153.

Moore, L. W. 1976. Crown gall, a "knotty" problem. Am. Rose Annu. 61:136-140.

Schroth, M. N., and Hildebrand, D. C. 1968. A chemotherapeutic treatment for selectively eradicating crown gall and olive knot neoplasms. Phytopathology 58:848-854.

Hairy Root

The pathogenic bacterium *Agrobacterium rhizogenes*, which causes hairy root, was reported by Hildebrand to infect roses in the northeastern United States in 1934 and again in 1937. The disease was of minor importance until 1952–1956, when serious losses were experienced in commercial rose production in southern California. Thereafter, the incidence of infection diminished and has remained insignificant.

Symptoms

Hairy root is characterized by swellings on stems or roots below the soil surface and usually at a wound such as disbud scars or the ends of cuttings. The swellings are firm but not woody as in crown gall, and discrete roots protrude from the swellings (Fig. 17). Large masses of fibrous roots 2–25 cm long are later formed. In storage under cool, moist conditions, many white roots are sometimes formed from the "hairy root" areas, which gives the disease one of its common names, "bristle root."

Usually, no characteristic symptoms appear on aerial plant parts. However, the second year after planting, infected plants have been observed to grow slowly, and many plants die. The death rate rises in the third and fourth years after planting.

Causal Organism

A. rhizogenes (Riker et al) Conn, the hairy root pathogen, was first named *Phytomonas rhizogenes* in 1930. The name was amended to *A. rhizogenes* in 1940.

Although *A. rhizogenes* is closely related to *A. tumefaciens*, the two bacterial pathogens have different infectious plasmids and the symptoms they cause in diseased plants are characteristically different. The virulence plasmids of *A. rhizogenes* are functionally similar to the Ti plasmids of *A. tumefaciens* but have diverged sufficiently to now represent a distinct plasmid type, based on DNA homology, compatibility, and virulence.

A simple bioassay for *A. rhizogenes* is to inoculate surface-sterilized carrot root slices in petri dishes with water infusions of infected roots. The infected carrot slices produce profuse roots in about 10 days.

Fig. 17. Hairy root caused by *Agrobacterium rhizogenes*. Note discrete fibrous roots protruding from swellings or galls.

The following hosts have been reported to be susceptible to isolates of *A. rhizogenes* from rose: *Beta vulgaris* L., *Chrysanthemum frutescens* L., *Coleus blumei* Benth., *Delphinium elatum* L., *Kalanchoë daigremontiana* Hamet. & Perrier, *Lonicera japonica aureo-reticulata* Nichols, *Lycopersicon esculentum* Mill. 'Pearson' and 'Earliana', *Malus sylvestris* Mill. (cultivar Delicious and a seedling selection), *Nicotiana tabacum* L., *Pelargonium hortorum* Bailey 'Penny', 'Irene', and 'Radio Red', *Phaseolus vulgaris* L. 'Blue Lake', *Pyrus communis* L. 'Bartlett', *Rosa* spp., *Rubus ursinus* var. *loganobaccus* Bailey, *Sedum spectabile* Boreau, and *Vicia faba* L. 'Windsor'.

The following plants are resistant when inoculated: *Begonia richmondensis* Hort., *Chrysanthemum morifolium* (Ramat.) Hemsl. 'White Top', *Cotoneaster parneyi* Hort., *Elaeagnus angustifolia* L., *Impatiens hostii* Engler & Warb., and *Spiraea vanhouttei* (Briot) Zabel.

Epidemiology
The development of hairy root is maximum when host growth is optimal. The disease is more severe at soil temperatures of 20°C than 26°C.

Control
Soil sterilization or fumigation, disinfestation of planting stock, and strict sanitation are important in controlling hairy root disease. Steam sterilization is an effective soil treatment. Cuttings can be disinfested by dipping them in 0.5% sodium hypochlorite; however, several precautions should be taken to prevent damage to cuttings during this treatment. Stems should not be allowed to dry out before they are dipped, so several cuttings should be made during a day and dipping should be avoided during periods of hot, dry weather. Tank solutions should be analyzed chemically each day, and additional hypochlorite should be added to maintain proper concentrations, since concentration of active material decreases with use and the pH rises rapidly. Commercial preparations of hypochlorite should be verified before they are used, because concentration decreases with time after bottling.

Selected References

Hildebrand, E. M. 1937. Infectious hairy root on rose. Plant Dis. Rep. 21:86-87.

Munnecke, D. E., Chandler, P. A, and Starr, M. P. 1963. Hairy root (*Agrobacterium rhizogenes*) of field roses. Phytopathology 53:788-799.

White, F. F., and Nester, E. W. 1980. Hairy root: Plasmid encodes virulence traits in *Agrobacterium rhizogenes*. J. Bacteriol. 141:1134-1141.

White, F. F., and Nester, E. W. 1980. Relationship of plasmids responsible for hairy root and crown gall tumorigenicity. J. Bacteriol. 144:710-720.

Diseases Caused by Viruses

Viral diseases are seldom lethal, but they generally reduce plant vigor, flower quality, and production yields significantly. One study estimated 14% losses in salable blooms from virus-infected glasshouse roses. A firm assessment of the economic advantages of growing roses that are free of viruses has not been possible as yet because of a lack of data on financial losses related to viral diseases and because, in general, knowledge of viral diseases of rose is somewhat limited. However, most major rose companies spend large amounts of money on virus-indexing procedures to ensure that rootstocks are free of known viruses.

In the past 20 years, several virus diseases of rose have been described. These viruses are differentiated primarily by symptom association and host range but purification, electron microscopy, comparison of physical properties, electrophoresis, serology, and other techniques may be used. The development of the enzyme-linked immunosorbent assay (ELISA) technique has provided a sensitive test for the presence of specific viruses in plant tissues. For the first time, we now have a method for detecting certain viruses in rose.

Rose Mosaic

Rose mosaic is found worldwide wherever roses are cultivated. It has been reported in the United States, England, New Zealand, Italy, Denmark, Germany, India, Australia, The Netherlands, South Africa, and Norfolk Island, South Pacific.

Symptoms
The symptoms associated with rose mosaic disease are highly variable. Characteristic symptoms include chlorotic line patterns, ring spots, and mottles in leaves sometime during the growing season (Plates 43–47). Yellow net and yellow mosaic symptoms are also associated with rose mosaic infections (Plates 48–50). No adverse effect on flower production has been reported, but foliar symptoms detract from the overall quality. Infected plants tend to be less vigorous than healthy plants and more sensitive to winterkill. Veinbanding may occur during prolonged periods at high temperatures (21°C or above).

Rose mosaic has been reported to occur particularly in cultivars of American origin. Symptoms are difficult to observe in *Rosa manetti* but are severe in Madame Butterfly, Ophelia, and Rapture cultivars. It is not unusual for infected plants to remain symptomless. The degree of symptom expression varies with the cultivar and the time of year as well as from year to year. In general, symptoms are most evident in spring.

Causal Agent
Rose mosaic virus (RMV), the causal agent of rose mosaic, has been generally associated with isolates of prunus necrotic ringspot virus (PNRSV). Similar symptoms on rose have been associated with single or mixed infections with PNRSV, apple mosaic virus (ApMV, recognized as a synonym for RMV by Fulton), and arabis mosaic virus (AMV). AMV, ApMV, and PNRSV separately or together in field-grown roses have

been reported to cause the range of symptoms recognized for rose mosaic. PNRSV alone induces chlorotic line patterns, ring spots, or leaf mottling; AMV plus PNRSV induces chlorotic veinbanding. Veinbanding may occur from PNRSV infections alone after prolonged periods of temperatures above 21°C.

Virions (virus particles) of these viruses are quasi-isometric and measure 25–29 nm for ApMV, 22–23 nm for PNRSV, and 30 nm for AMV. Negatively stained particles in electron micrographs show some oval and bacilliform particles. Thermal inactivation points (10 minutes) are 54°C, 55–62°C, and 55–61°C for ApMV, PNRSV, and AMV, respectively.

There is no evidence for vector transmission of viruses in rose, although AMV is known to be transmitted by the free-living, soil-inhabiting nematodes *Xiphinema diversicaudatum* and *X. coxi*. Virus transmission in rose appears to be limited to vegetative propagation when virus-infected buds, scions, or rootstocks are grafted to healthy plants. PNRSV is pollen-transmitted in fruit trees. Pollen transmission is suspected to occur in roses also, since spread in the field is slow.

The ELISA test is reliable for detecting ApMV and PNRSV in rose. Both viruses are detected as accurately with antisera of ApMV and PNRSV in combination coated to ELISA plates as with ApMV and PNRSV antisera coated individually to plates. Detection of ApMV and PNRSV is usually best from petal tissue and leaf specimens in spring and autumn rather than in midsummer. Sampling young succulent leaf tissue is necessary to detect the viruses in foliage. AMV has occasionally been undetectable in rose tissues with ELISA. Serologically specific electron microscopy (SSEM) is very sensitive for detecting AMV and is about twice as sensitive as ELISA for detecting all three viruses in rose extracts.

Host Range
Diagnostic species: PNRSV—*Cucumis sativus, Momordica balsamina, Cyamopsis tetragonoloba, Prunus serrulata* 'Shirofugen'. ApMV—*Cucumis sativus, Torenia fournieri, Catharanthus roseus, Vigna sinensis, Malus sylvestris*. AMV—*Chenopodium amaranticolor, C. quinoa, Cucumis sativus, Nicotiana tabacum* 'White Burley', *Phaseolus vulgaris, Petunia hybrida*.

Propagation species: PNRSV—*Prunus persica, P. mahaleb, Catharanthus roseus, Chenopodium quinoa*. ApMV—*Catharanthus roseus, Chaenomeles japonica, Chenopodium quinoa, Cucumis sativus*. AMV—*Petunia hybrida, Nicotiana clevelandii*.

Assay species: PNRSV—*Momordica balsamina, Cucumis sativus, Chenopodium quinoa*. ApMV—*Cyamopsis tetragonoloba, Chenopodium quinoa, Cucumis sativus*. AMV—*Chenopodium amaranticolor*.

General host range: PNRSV—Fairly wide host range; known to infect species in 21 dicotyledonous families. ApMV—Wide host range; known to infect more than 65 species in 19 families. AMV—Wide host range; known to infect 93 species in 28 dicotyledonous families; also reported to infect the roots of the gymnosperm *Chamaecyparis lawsoniana*.

Control
Rose mosaic is controlled by removing infected plants and by using virus-indexed propagative plants. Madame Butterfly, Ophelia, and Rapture show severe symptoms and thus are useful indicator cultivars for bioassays by graft transmissions. Flowering cherry, *Prunus serrulata* 'Shirofugen', reacts to grafts of virus-infected buds with a distinctive local hypersensitive reaction of necrosis and gumming around the graft site 30 days after budding. Positive reactions are best from basal branches.

Heat treatment of infected plants can be used to obtain buds free of RMV. Buds taken from plants held at 38°C for four weeks are usually RMV-free and can be used to propagate "clean" cultivars.

Selected References

Barbara, D. J. 1981. Detecting prunus necrotic ringspot virus in Rosaceous hosts by enzyme linked immunosorbent assay. Acta Hortic. 94:329-332.

Basit, A. A., and Francki, R. I. B. 1970. Some properties of rose mosaic virus from South Australia. Aust. J. Biol. Sci. 23:1197-1206.

Casper, R. 1973. Serological properties of prunus necrotic ringspot and apple mosaic virus isolates from rose. Phytopathology 63:238-240.

Fleisher, Z., Drori, T., and Loebenstein, G. 1971. Evaluation of Shirofugen flowering cherry as a reliable indicator for rose mosaic virus. Plant Dis. Rep. 55:431-433.

Secor, G. A., Kong, M., and Nyland, G. 1977. Rose virus and virus-like diseases. Calif. Agric. 31(3):4-7.

Thomas, B. J. 1980. The detection by serological methods of viruses infecting rose. Ann. Appl. Biol. 94:91-101.

Thomas, B. J. 1981. Studies on rose mosaic disease in field-grown roses produced in the United Kingdom. Ann. Appl. Biol. 98:419-429.

Strawberry Latent Ringspot Virus

Strawberry latent ringspot virus (SLRV) occurs naturally in many species of wild and cultivated plants. Of 167 species of dicotyledonous plants inoculated mechanically, SLRV infects 126 belonging to 27 families. Many species may be infected systemically without showing symptoms. SLRV is frequently found in England and is thought to be the most economically important viral disease of rose in that country.

Symptoms
Characteristic symptoms are small, angular yellow flecks on leaves and a marked stunting of shoots and leaves. Leaves may have a leathery texture and appear distorted or strapped.

Some cultivars show no symptoms. Symptoms vary in severity, depending on cultivar response, environmental factors (e.g., temperature), mixed infections with arabis mosaic virus (AMV) and prunus necrotic ringspot virus (PNRSV), and differences in virulence of SLRV isolates. Symptoms are most intense at 12°C and are masked at 23°C.

Causal Agent
SLRV particles are isometric and about 30 nm in diameter. Isometric particles are frequently in tubular sheaths. Infectivity is lost after 10 minutes at 52–58°C or after dilution to 10^{-3} to 10^{-5} of *Chenopodium quinoa* sap.

SLRV is transmitted by the nematode *Xiphinema*

diversicaudatum; however, the importance of this vector as a mechanism of dispersal is uncertain in rose production. The nematode has not been consistently associated with diseased plants, and infected rootstocks are proposed to be the only significant source of infection.

The ELISA test is reliable for detecting SLRV in rose, while serologically specific electron microscopy (SSEM) is more sensitive for detecting the virus in rose extracts. As little as 0.5 µg/ml can be detected in leaf extracts from *Rosa multiflora* with SSEM.

Chenopodium amaranticolor, C. quinoa, and *Cucumis sativus* 'Butcher's Disease Resister' are good bioassay plants. Bioassay is most efficient from young infected rose leaves, and mechanical transmission from rose is improved when polyvinylpyrrolidone is used in the extraction buffer.

Host Range
Diagnostic species: *Chenopodium amaranticolor, C. murale, C. quinoa, Cucumis sativus, Nicotiana rustica, N. tabacum, Petunia hybrida.*

Propagation species: *Cucumis sativus.*

Assay species: *Chenopodium murale, C. amaranticolor.*

Control
SLRV is controlled by removing infected plants, by using virus-indexed rootstocks, and by sterilizing soil to reduce the population of nematodes that may potentially act as vectors.

Selected References

Itkin, R., and Frost, R. R. 1974. Virus diseases of roses. I. Their occurrence in the United Kingdom. Phytopathol. Z. 79:160-168.

Itkin, R., and Frost, R. R. 1976. Virus diseases of roses. II. Strawberry latent ringspot virus. Phytopathol. Z. 87:205-223.

Thomas, B. J. 1980. The detection by serological methods of viruses infecting the rose. Ann. Appl. Biol. 94:91-101.

Thomas, B. J. 1981. Detection by serologically specific electron microscopy of viruses infecting rose and other hardy nursery stock plants. Micron 12:175-176.

Rose Streak

Rose streak occurs in Europe and in the eastern United States and has been found infrequently in California. The disease is suspected to be caused by a virus, called rose streak virus (RSV).

Symptoms
Characteristic symptoms of rose streak include brownish green rings and veinbanding in expanded leaves; leaves drop prematurely. The foliar symptoms may be accompanied by ring patterns on stems and sometimes on fruit.

Types of rose cultivars that may show symptoms from infection include teas, hybrid teas, hybrid perpetuals, hybrid multifloras, hybrid wichuraianas, hybrid rugosas, hybrid Bengals, Noisettes, Chinas, and polyanthas. Symptoms on rugosas are generally mild or undetectable, whereas hybrid multifloras and wichuraianas are more susceptible and show severe symptoms.

Causal Agent
Little is known about RSV other than that it is graft-transmitted and appears to affect only roses. Madame Butterfly, Ophelia, Rapture, and Briarcliff cultivars are sensitive indicators of RSV. Necrosis and blackening develop around RSV-infected buds grafted to these cultivars within two months after the union has been established.

Control
Field spread of rose streak is believed to occur only from infected propagation material. Control is based on removing and destroying infected plants.

Selected Reference

Secor, G. A., Kong, M., and Nyland, G. 1977. Rose virus and virus-like diseases. Calif. Agric. 31(3):4-7.

Rose Rosette

Rose rosette occurs as a natural infection on wild rose species in northeastern California and has been experimentally transmitted to cultivated roses through grafts. The disease appears to be endemic in the midwestern United States in Kansas, Nebraska, and Missouri, where *Rosa multiflora* hedges are frequently planted and become infected. Spread of the disease to cultivated roses is fairly common in this area.

Symptoms of rose rosette in *R. multiflora* 'Burr' are leaflet distortion and wrinkling, bright red leaf pigmentation, witches'-broom, and phyllody. Affected canes are often excessively thorny and slow to mature. The plant usually dies as the symptoms spread and eventually affect all canes.

The causal agent of rose rosette is thought to be transmitted by mites, but the viral etiology has not yet been established.

Selected Reference

Secor, G. A., Kong, M., and Nyland, G. 1977. Rose virus and virus-like diseases. Calif. Agric. 31(3):4-7.

Rose Ring Pattern

Rose ring pattern occurs in commercially grown roses in California and Oregon. It was first observed in 1973 in *Rosa multiflora* 'Burr', which serves as a reliable indicator for the causal agent (Plate 51).

Symptoms occur on new growth that is forced by removing the leaves 10 days after grafting. Four weeks later, new leaflets appear stunted, distorted, rugose, and mottled (Plate 52). The disease has been readily graft-transmitted to numerous cultivars, to *R. rugosa*, and to major rootstocks. Most cultivars react to infection with symptoms consisting of rings, fine line patterns, and chlorotic flecking of the leaves (Plate 53). Symptoms may resemble those caused by rose mosaic virus (RMV), but rose ring pattern does not cause a necrotic reaction in Shirofugen cherry. Symptoms in cultivar Queen Anne are striking yellow blotches on leaflets and color breaking in ring patterns in petals. Madame Butterfly also exhibits color break symptoms (Plate 54). Some rootstocks may remain symptomless when infected; however, faint ring or line patterns may develop in *R. manetti, R. odorata* cv. Sweet, and Dr. Huey.

The virus or viruslike causal agent(s) of rose ring pattern has not yet been identified. The causal agent appears to be disseminated through propagative plant material; there is no evidence for natural spread in field or glasshouse roses. The causal agent is sensitive to thermal therapy; buds free of rose ring pattern are obtained from infected plants treated three to four weeks at 38°C.

Rose ring pattern is controlled by removing and destroying diseased plants and by using healthy rootstocks and buds. Remission of symptoms in *R. multiflora* is obtained after treatments with the antiviral compound ribavirin.

Selected Reference

Secor, G. A., and Nyland, G. 1978. Rose ring pattern: A component of the rose-mosaic complex. Phytopathology 68:1005-1010.

Rose Wilt

Rose wilt was first reported in New Zealand and Australia in 1931. The disease has since been reported in California. The identity of the causal agent is unknown.

Symptoms

Symptoms are influenced by cultivar, age of plant at the time of infection, and the environment. Characteristic symptoms are downward curling of leaves, extensive veinclearing, epinasty, and premature leaf abscission. Shoots are often weak at the point of graft attachment. Cultivars with green foliar pigmentation tend toward chlorosis and those with red pigmentation tend toward dull coloration. Established plants develop general decline, extensive dieback, and loss of apical dominance. Leaf balling occurs in spring, along with rosetted growth and small, incurved leaves. Plants show reduced vigor and extensive shoot dieback and ultimately die.

Rose wilt has been reported to be confused with wilt caused by *Verticillium albo-atrum*; however, a basic difference is that *Verticillium* causes wilting of leaves at the tips of young canes and yellowing of lower canes. Furthermore, *Verticillium* has not been isolated from plants with the symptoms associated with rose wilt.

Control

The practice of obtaining understock cuttings from heads of previously budded stock is discouraged. Stock beds of *Rosa multiflora* known to be free of known viruses should be used. Clonal selections should be made carefully and early and not at the time of budwood collection. Diseased plants should be removed and destroyed.

Selected References

Fry, P. R., and Hammett, K. R. W. 1971. Rose wilt virus in New Zealand. N.Z. J. Agric. Res. 55:431-433.

Grieve, B. J. 1931. Rose wilt and dieback. A virus disease of roses occurring in Australia. Aust. J. Exp. Biol. Med. Sci. 8:107-121.

Hammett, K. R. W. 1971. Symptom differences between rose wilt virus and *Verticillium dahliae* wilt of roses. Plant Dis. Rep. 55:916-920.

Rose Spring Dwarf

Rose spring dwarf is found in commercial nurseries, landscape roses, and public rose gardens in California. Characteristic symptoms, produced when leaves first emerge in the spring, include rosetting or a "balled" appearance in the new growth following bud break (Plate 55). The leaves are recurved or very short, are borne on arrested shoots (Plate 56), and show conspicuous veinclearing or a netted appearance. Shoot elongation may be delayed for two months, and newly developing leaves may be symptomless. Infected plants may remain symptomless during the summer and fall, but plant vigor is reduced and some mild epinasty, veinclearing, or both may show on new growth in the fall. Branches often develop a zigzag growth appearance (Plate 57).

Rosa multiflora cv. Burr and Dr. Huey show symptoms well. Cultivar Burr is useful as an indicator plant (Plate 58); bud chips are grafted at several locations along the cane, and symptoms may be induced by stripping plants of young foliage and pruning to mature buds. Characteristic rosetting and veinclearing appear in two to three weeks and necrotic pitting of the xylem after a year of growth.

The causal agent of rose spring dwarf is unknown and does not appear to spread in nature. The disease is not sensitive to temperatures used for thermal therapy. The best control measure is to rogue infected plants.

Selected References

Slack, S. A., Traylor, J. A., Nyland, G., and Williams, H. E. 1976. Symptoms, indexing, and transmission of rose spring dwarf disease. Plant Dis. Rep. 60:183-187.

Traylor, J. A., Wagnon, H. K., and Williams, H. E. 1971. Rose spring dwarf, a graft-transmissible disease. Plant Dis. Rep. 55:294.

Rose Leaf Curl

Rose leaf curl is widely distributed throughout the United States in or near "antique" roses in public rose gardens. Characteristic symptoms resembling rose wilt (see Rose Wilt) occur on hybrid tea roses but not on rootstock cultivars. Symptoms are first seen in the spring as reduced leaf size, easily detached leaflets, leaf epinasty, necrosis of shoot tips (Plates 59 and 60), and a yellow flecking of veins (Plate 61) that may progress into necrosis. Shoots are characteristically pointed with a broad base. Plants may recover during the summer, but symptoms reappear in the fall, usually as leaf epinasty and as cracking, internal necrosis, longitudinal corky areas, and xylem pitting in mature canes (Plate 62).

The causal agent of rose leaf curl has not been determined. Rose cultivars Queen Elizabeth and Madame Butterfly are good indicator plants. Recommended control measures are to remove affected plants and destroy them, since natural spread of this disease is very slow.

Selected References

Secor, G. A., Kong, M., and Nyland, G. 1977. Rose virus and virus-like diseases. Calif. Agric. 31(3):4-7.

Slack, S. A., Traylor, J. A., Williams, H. E., and Nyland, G. 1976. Rose leaf curl, a distinct component of a disease complex which resembles rose wilt. Plant Dis. Rep. 60:178-182.

Rose Flower Break

Rose flower break occurs in England, New Zealand, and Australia. Little is known about the causal agent and its economic importance, but the disease is a potential threat because flower quality is severely reduced. Characteristic symptoms are distortion of flower petal margins and intense color development in the veins of petals.

The disease is transmitted by grafts with buds, petals, or leaf tissue. Although virus particles have been detected in affected flowers, their association with the disease has not yet been established.

Rose flower break is controlled by removing and destroying affected plants when the disease is found.

Selected References

Farrar, E. H., and Frost, R. R. 1972. A new disease of roses in Britain. Plant Pathol. 21:97.

Hunter, J. A. 1966. Rose colour break: A virus disease of roses. N.Z. J. Agric. Res. 9:1070-1072.

Itkin, R., and Frost, R. R. 1974. Virus diseases of roses. I. Their occurrence in the United Kingdom. Phytopathol. Z. 79:160-168.

Rose Flower Proliferation

A flower proliferation, in which the center of the flower elongates into a stem that bears additional flowers, has been described in Italy. A similar response can occur when some cultivars are grown under high nitrogen fertilization. Little is known about this disease, although another flower anomaly, in which petals turn green and leaflike, has been described in England. The cause of these diseases is not known, nor is it known if they are transmissible.

Selected References

Gualaccini, F. 1963. Singolare vineacenza, prolungamento dell'asse fiorale e anomalie fogliari su Rosa ottenuta da seme di plants affetta da' mosaico giallo ('Rose Yellow Mosaic'). Boll. Stn. Patol. Veg., Rome 21:45-46.

Itkin, R., and Frost, R. R. 1974. Virus diseases of roses. I. Their occurrence in the United Kingdom. Phytopathol. Z. 79:160-168.

Tobacco Streak Virus

Tobacco streak virus is found in *Rosa setigera*. Symptoms include irregular chlorotic areas, vein chlorosis, and twisting leaves. The virus has been found in wild rose from several collecting sites in Oregon. The incidence of tobacco streak virus is low in rose cultivars; the symptoms are more severe than those caused by rose mosaic virus, and it is unlikely that infected plants would be selected for propagation.

Selected References

Converse, R. H., and Bartlett, A. B. 1979. Occurrence of viruses in some wild *Rubus* and *Rosa* species in Oregon, U.S.A. Plant Dis. Rep. 63:441-444.

Fulton, R. W. 1970. A disease of rose caused by tobacco streak virus. Plant Dis. Rep. 54:949-951.

Diseases Caused by Nematodes

Plant-parasitic nematodes are a problem on roses worldwide. Nematode-associated rose problems have been reported in widely separated locations, including all parts of the United States, England, Denmark, Japan, and India. This may well be expected, because plant-parasitic nematodes have broad, nearly worldwide distribution, and many (especially root-knot nematodes) have a broad host range and are known to attack nearly all cultivated plants. Furthermore, plant-parasitic nematodes are readily spread worldwide by commercial production practices used for propagating roses. Root-lesion and root-knot nematodes are particularly troublesome, because they are endoparasitic and are easily distributed with rose rootstocks.

The magnitude of nematode problems of roses is not easily assessed because the symptoms nematodes cause—reduced plant growth, chlorosis, and reduced flowering—may also be caused by other factors that reduce and interfere with adequate, properly functioning plant root systems. Furthermore, rose rootstocks vary in sensitivity to nematode invasion. Tolerance of nematode infestation is only one of the characteristics for which rose rootstocks are selected; compatibility with the flowering cultivar, influence on length of flowering period, resistance to *Verticillium*, and tolerance of a restricted soil volume must also be considered when rootstocks are chosen.

Symptoms

Symptoms on aerial plant parts include reduced vigor, dwarfed leaves and shoots, chlorosis, wilting, leaf drop, reduced flower quality (stem length and flower size), and increased susceptibility to root rot pathogens. These symptoms are characterized as a general decline (Plate 63). Plants showing decline symptoms are usually discarded because of decreased flower production and poorer flower quality. Frequently, plants replanted into areas that have been rogued because of nematode infestations again show symptoms after a year or two.

Several different symptoms may occur on roots, depending on the nematode species and the number of feeding nematodes. Clear-cut symptoms may rarely be seen, and roots may show gradations of a number of symptom types. Therefore, positive diagnosis of nematodes as the cause of observed symptoms requires recovery of nematodes from the roots and/or soil surrounding symptomatic plants, followed by a laboratory identification. However, symptoms that are usually associated with specific nematodes and that are useful as a general guideline are shown in Table 1. With the exception of root galls (Plates 64–66), these symptoms are not specific to nematode infestation. Moreover, *Xiphinema* and *Meloidogyne* both may produce root galls on roses. Based on symptoms alone, *X. diversicaudatum* may be easily mistaken for *M. hapla*; however, the trained diagnostician can distinguish between them.

Causal Organisms

Nematodes that have been found on roses include *Xiphinema* (dagger), *Paratylenchus* (pin), *Macroposthonia* (*Criconemoides*) (ring), *Pratylenchus* (lesion), *Helicotylenchus* (spiral), *Hemicycliophora* (sheath), *Belonolaimus* (sting), *Trichodorus* (stubby root), *Meloidogyne* (root-knot), *Ditylenchus* (stem and bulb), *Aphelenchoides* (leaf and stem), *Psilenchus*, *Tylenchus*, *Tylenchorhynchus*, *Hoplolaimus*, *Longidorus*, *Rotylenchus*, and *Neotylenchus*. The species most frequently found associated with roses are *X. diversicaudatum* (Micoletzky) Thorne, *Helicotylenchus nannus* Steiner, *Pratylenchus penetrans* Cobb, *Pratylenchus vulnus* Allen & Jensen, *Meloidogyne hapla* Chitwood, and *Macroposthonia axeste* Fassuliotis & Williamson.

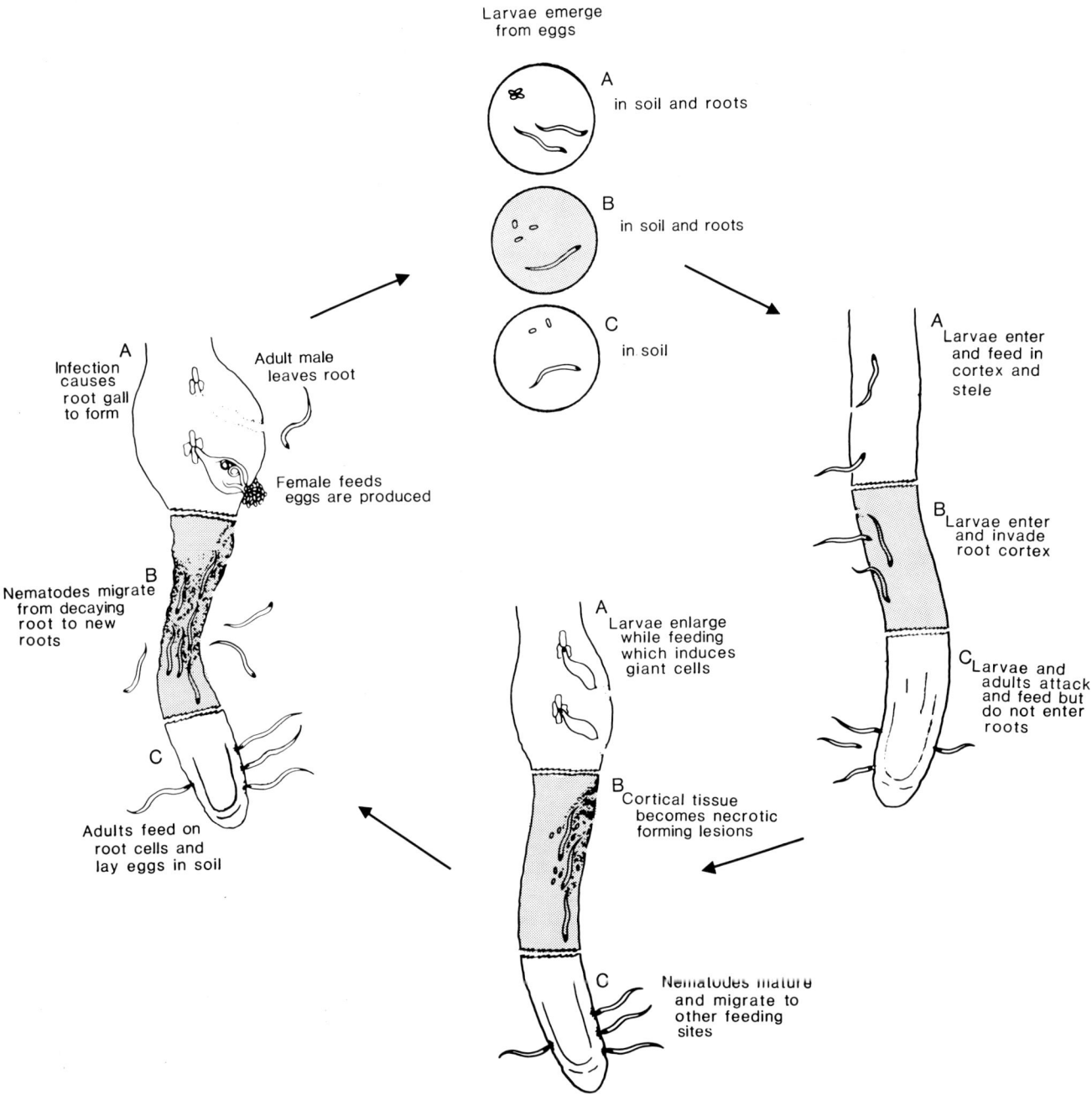

Fig. 18. Life cycles of *Meloidogyne* sp. (A), *Pratylenchus* sp. (B), and *Xiphinema* sp. (C). These nematodes are shown in one diagram for convenience. Although they usually infest different plants, they may in some instances be found in the same root but not necessarily in the same spatial orientation shown in the diagram.

Nematodes may be found on roots of glasshouse roses as well as field-grown plants and on dormant or bare-root plants. A general diagram of the life cycles of the nematode genera most commonly found to infect roses is shown in Figure 18.

M. hapla invasion and feeding in roots cause formation of giant cells, hyperplasia of cortical and vascular parenchyma, formation of xylem elements from vascular parenchyma, and retardation of meristematic activity in root tips. Gall formation from *X. diversicaudatum* infestation is caused by a hyperplastic response of cortical cells. Cells in feeding sites increase two to three times in size. Meristematic activity in infested root tips is retarded, and vascular differentiation extends far into root tips.

Control

Measures for controlling nematodes on roses include fumigating field nurseries, fumigating or steam-sterilizing glasshouse beds before planting, restricting the movement of infected stock plants, and applying postplanting nematicides. Treatments for eliminating nematodes in raised or watertight ground beds are very effective; however, in open ground beds, plant roots penetrate deeply into the soil and associated plant-parasitic nematodes follow the root system (Plate 67). This presents a problem in getting the treatment to the area of the soil where it can be effective.

Hot-water treatment effectively controls nematodes on planting material, but such treatments are often damaging to rose roots. Thermal injury to roses can be reduced by pretreating plants at 38°C for 24 hours to induce heat-hardening. Nematode infestations can then be eliminated by partially immersing the rose plants at 48°C for 35 minutes.

TABLE 1. Symptoms Associated with Nematodes on Roses

Genus	Common Name	Symptoms
Xiphinema	Dagger	Root galls at tip of feeder roots; swelling and curling tips, called "curly tip"
Meloidogyne	Root-knot	Galls on smaller roots; *M. hapla* causes excessive root branching
Pratylenchus	Lesion	Root lesions that become necrotic
Macroposthonia	Ring	Root lesions that become necrotic
Rotylenchus	Spiral	General surface browning or root discoloration
Helicotylenchus	Spiral	General surface browning or root discoloration
Tylenchorhynchus	Stylet (stunt)	Injured root tips

Rose rootstocks have been tested for their suitability as hosts for various parasitic nematodes. Most rootstocks are hosts for both *Pratylenchus penetrans* and *P. vulnus*; *Rosa indica* 'Major' and *R. multiflora* '60-5' are very susceptible. *M. hapla* reproduces well on *R. indica* 'Major' but not on *R. noisettiana* 'Manetti'; thus, Manetti appears to be highly resistant to *M. hapla*. Dr. Huey, Manetti, *R. odorata*, and other *Rosa* spp. are good hosts for *P. vulnus*, whereas *R. multiflora* is less susceptible.

Selected References

Collen, W. A., and Hendricks, G. J. 1972. Investigations on the resistance of rose rootstocks to *Meloidogyne hapla* and *Pratylenchus penetrans*. Nematologica 18:155-158.

Davis, R. A. 1960. Nematodes associated with roses. Am. Rose Annu. 45:34-47.

Hart, W. H., and Maggenti, A. R. 1971. Control of root-knot nematode, *Meloidogyne hapla*, on 2-year-old field-grown roses, *Rosa multiflora japonica*. Plant Dis. Rep. 55:89-92.

Hayashi, I. 1976. Studies on parasitic nematodes and their control in greenhouse roses. Bull. Kanagawa Hortic. Exp. Stn. 23:54-63.

Jackobsen, J. 1975. Plant parasitic nematodes on roses in Danish glasshouses. Tidsskr. Planteavl 79:489-494.

Muthukrishnan, T. S., Chandrasekaran, J., Lakshmanan, P., and Chinnarajan, A. M. 1975. Occurrence of the nematode *Hemicycliophora labiata* Colbran 1960 on roses in Tamil Nadu. South Indian Hortic. 23:75-76.

Ohkawa, K., and Saigusa, T. 1981. Resistance of rose rootstocks to *Meloidogyne hapla*, *Pratylenchus penetrans*, and *Pratylenchus vulnus*. HortScience 16:559-560.

Santo, G. S., and Lear, B. 1976. Influence of *Pratylenchus vulnus* and *Meloidogyne hapla* on the growth of rootstocks of rose. J. Nematol. 8:18-23.

Schindler, A. F. 1956. Nematodes associated with roses in a survey of commercial greenhouses. Plant Dis. Rep. 40:277-278.

Towson, A. J. 1982. The role of heat-hardening and dormancy in increasing tolerance of hot-water treatment of rose to control endoparasitic nematodes. Ph.D. thesis. University of California, Davis. 90 pp.

Towson, A. J., and Lear, B. 1981. Injury to bare rooted roses hot water treated for elimination of endoparasitic nematodes. J. Nematol. 13:462.

Williamson, C. E. 1969. Nematode problems. Pages 211-219 in: Roses: A Manual on the Culture, Management, Diseases, Insects, Economics and Breeding of Greenhouse Roses. J. W. Mastalerz and R. W. Langhans, eds. Penn. Flower Growers, N.Y. State Flower Growers Assoc., Inc., and Roses Inc. 331 pp.

Winfield, A. L. 1974. Observations on the occurrence, pathogenicity and control of *Pratylenchus vulnus*, *Pratylenchus thornei*, and *Xiphinema diversicaudatum* associated with glasshouse roses. Ann. Appl. Biol. 77:297-307.

Winfield, A. L. 1974. Damage to roses by *Longidorus macrosoma*. Plant Dis. Rep. 58:913-914.

Part II. Noninfectious Diseases[1]

Physiologic Problems

Blindness

Flower formation in greenhouse roses is an autonomous, self-induced function and does not depend on external environmental factors such as specific vernalizing temperatures or inductive daylengths. Floral initiation begins in an actively growing meristem within 14–16 days of the removal of the previous marketable stem. New flowers form on new shoots arising from axillary buds on basal stem tissue that contains at least two five-leaflet leaves and that was left on the plant at the time of harvest. At this time, the newly emerged axillary shoots are about 5 cm long.

Normally, each developing shoot would terminate in a flower. However, in some cultivars and under some conditions, flower buds either 1) fail to form at all, 2) begin to form but then abort, along with the uppermost few leaves, or 3) atrophy and abscise. The resultant shoots are termed "blind shoots" (Plate 68).

Blind shoots remain thin and short, bear fewer leaves than flowering shoots, elongate very slowly, and may remain quiescent for long periods. The leaves of flowering shoots contain higher levels of chlorophyll and anthocyanin pigments than the leaves of blind shoots.

Shoots originating from axillary buds below the top one on any particular branch are much more likely to be blind than the shoot growing above them. Blindness is more likely to occur during periods of low light and temperature than under brighter, warmer conditions. Cultivars that show high percentages of blindness when grown at 12°C may develop almost all flowering shoots if grown at 18–24°C. Light and temperature affect the concentrations and distribution of endogenous plant hormones within rose shoots. The light and temperature levels favorable for flowering are well correlated with the levels of gibberellins and auxins in the young developing leaves on the stem. If conditions are such that the production of these growth-promoting substances is low, the initiating flower bud is likely to atrophy or abort during an early stage of its development and before the initiation of anthers and pistils. If floral initiation progresses beyond the formation of anthers and pistils, these organs begin to synthesize their own gibberellins and they become self-sufficient and no longer dependent on other plant parts (the young leaves) for their supply of these hormones.

Conditions that favor blindness—low light levels, low temperatures, etc.—also favor the production of above-normal quantities of ethylene by the young shoots, and again excellent correlations exist between the degree of blindness and a balance between the total level of metabolic activity and the levels of ethylene generated.

It now appears that blindness in roses is related to a number of critical metabolic processes going on within the plant, the most decisive of these processes being competition for a limited supply of photosynthates and other metabolites. This competition is affected to a considerable extent by apical dominance, which favors the upper shoot on a stem on which more than the uppermost bud has broken into growth. Gibberellins are assumed to encourage flower development by causing the developing bud to become a more powerful metabolic sink for the raw materials it needs for continuing nourishment and development.

Selected References

Horridge, J. S., and Cockshull, K. E. 1974. Flower initiation and development in the glasshouse rose. Scientia Hortic. 2:273-284.

Moe, R. 1971. The relationship between flower abortion and endogenous auxin content in rose shoots. Physiol. Plant. 24:374-379.

Nell, T. A., and Rasmussen, H. P. 1979. Floral development and blindness in roses: An SEM study. J. Am. Soc. Hortic. Sci. 104:18-20.

Nell, T. A., and Rasmussen, H. P. 1979. Blindness in roses: Effects of high intensity light and blind shoot prediction techniques. J. Am. Soc. Hortic. Sci. 104:21-25.

Zieslin, N., and Halevy, A. H. 1976. Flower bud atrophy in Baccara roses. VI. The effect of environmental factors on gibberellin activity and ethylene production in flowering and non-flowering plants. Physiol. Plant. 37:331-335.

Bullheads

One of the malformations that occur in roses, known as "bullheading," is prevalent in cultivars such as Baccara and Talisman and in cultivars belonging to the Columbia family of roses, which were popular during the 1930–1950 period. Bullheads are characterized by 1) a lower than normal ratio of length to diameter of the flower bud, giving a flat-topped appearance to the bud instead of the pointed tip normally preferred; 2) an increase in size and weight of the flower bud; 3) an increase in the number of short petals and petaloids; and 4) the proliferation of secondary florets bearing mainly

---

[1]Part II was prepared by Marlin N. Rogers.

carpels near the base of the flower.

Bullheads are more prevalent when temperatures drop to low levels (12–15°C) during early flower development, and they are correlated with marked changes in the balance of different growth regulators present within the flowers. Low temperatures appear to reduce gibberellin activity and increase cytokinin activity, which results in proliferation in the nectary area and promotes the development of the adventitious florets that are associated with bullheads.

The problem has been alleviated experimentally by the injection of gibberellins into the receptacle of flowers being grown under low-temperature regimes, before the flower buds reach a diameter of 9 mm. Growing the plants at 18–24°C essentially eliminated bullheading in Garnette and Zorina cultivars and substantially reduced it in Baccara.

Selected References

Moe, R. 1971. Factors affecting flower abortion and malformation in roses. Physiol. Plant. 24:291-300.

Zieslin, N., Madori, G., and Halevy, A. H. 1979. Involvement of hormonal balance in the control of the 'bullhead' malformation in Baccara rose flowers. J. Exp. Bot. 30:15-25.

Bent Neck

A frequently observed problem with cut roses is premature death caused by a loss of rigidity in the pedicel a few centimeters below the opening flower bud, causing the symptom called "bent neck" or "rose neck droop." When cells in the neck area lose turgor, the neck bends or droops from a lack of mechanical strengthening tissues, such as lignified water-conducting vessels or collenchymatous tissues present within the stem at that point. This weakness has been shown to be essentially a water balance problem, in which the rate of water loss from the rose exceeds the rate of water uptake, with the resulting loss of turgor of the cells in the pedicel just below the flower. Research has shown that the most common cause of bent neck is an impeded flow of water through the stem of the cut rose, due to air blockage of the water-conducting vessels, physiologic plugging, or direct or indirect microbial plugging of the water-conducting system.

Many different postharvest handling procedures have been shown to help prevent the problem, including recutting rose stems under water, using solution temperatures of 40°C, using gasfree holding solutions, using solutions containing antimicrobial chemicals, and using wetting agents in the holding solutions to reduce surface tension and enhance water movement through the water-conducting system.

Selected Reference

Burdette, A. N. 1970. The cause of bent neck in cut roses. J. Am. Soc. Hortic. Sci. 95:427-431.

Environmental Imbalances

Petal Blackening in Baccara Roses

Blackening of the petals of Baccara roses grown outdoors or in unheated greenhouses in Israel was not observed in flowers produced at the same time in heated greenhouses. Study of the problem showed that the concentrations of the two red pigments, cyanidin and pelargonidin, found in this cultivar were considerably higher in the black flowers than in the normal red flowers. High polyphenolase activity and high concentrations of tannins were also found in the black flowers but not in the red flowers. The blackening phenomenon was attributed to the increase in pigment content at low temperatures and to the accumulation of oxidation products of polyphenols.

Selected Reference

Zieslin, N., and Halevy, A. H. 1969. Petal blackening in 'Baccara' roses. J. Am. Soc. Hortic. Sci. 94:629-631.

Petal Bluing in Baccara Roses

Under some conditions, flower buds of Baccara roses show an unattractive and undesirable bluish coloration instead of the bright, clear red normally expected. Studies in Israel have shown that bluing may occur in different situations with a variety of causes.

Bluing in senescing petals was accompanied by increasing pH levels and decreasing levels of malic acid in the petal tissue, resulting in copigmentation (the association of anthocyanins with other flavonoids and related compounds). Bluing in young flower petals was not accompanied by similar changes but by a decrease in the concentration of cyanadin and pelargonidin (both per unit weight and per unit of petal area), which resulted in the appearance of bluing. In still other cases, factors that caused an increase in the cyanin-pelargonin ratio with no significant change in total pigment content also resulted in increased bluing.

Most of the growth and enlargement of rose flower buds occurs during a short period of time after the flowering shoot ceases elongating, and 90% of the synthesis of the flower pigments occurs during the final expansion of the bud. Environmental stresses such as high temperatures or low light intensity during this short interval were found to reduce pigment formation, which resulted in more bluing of the petals. These environmental effects work by affecting the availability of photosynthates accumulating in the flower bud. High temperatures, low light intensity, and low levels of carbon dioxide in the atmosphere caused bluing, whereas the opposite environmental conditions prevented bluing. Injections of gibberellic acid into the receptacle of the flower bud also enhanced red pigmentation, suggesting that the acid acted to intensify the strength of the flower bud as a metabolic sink for photosynthates being manufactured by the leaves.

Selected References

Asen, S. 1975. Factors affecting flower colour. Acta Hortic. 41:57-68.

Biran, I., Enoch, H. Z., Zieslin, N., and Halevy, A. H. 1973. The influence of light intensity, temperature and carbon dioxide concentration on the anthocyanin content and blueing of 'Baccara' roses. Scientia Hortic. 1:117-164.

Biran, I., Robinson, M., and Halevy, A. H. 1974. Factors determining petal color of Baccara roses. III. Effect of the ratio between cyanin and pelargonin. J. Exp. Bot. 25:632-637.

Heat and Moisture Stress

Heat stress, moisture stress, and/or salinity may cause almost identical symptoms on rose plants, and all three are often interrelated. A concentration of soluble salts that might cause no injury symptoms when the growing medium is kept moist and when low levels of light and temperature reduce transpirational water losses might cause severe scorch (marginal necrosis of stems and leaves) under high light and temperatures or when the growing medium dries out (Plate 83). Soil moisture and salinity levels affect the rate of water uptake, and aerial light and temperature levels affect the rate of water loss (transpiration) from the plant. The scorch comes about as a result of the unfavorable water balance conditions within the leaves, with the cells around the edges of the leaf becoming so desiccated that permanent injury and necrosis occur. The necrosis is worst of all when bright, warm, sunny days follow a period of dark, cool, cloudy weather that promoted the development of soft, succulent top growth especially susceptible to injury.

Oxygen Deficiency

Roses grown in poorly structured growing media with poor drainage, or in media maintained at excessively high moisture levels, display foliar symptoms that have been shown to be caused by oxygen deficiency (Plate 72). Soil aeration is insufficient to supply the oxygen the roots need to carry on normal respiratory activities. The foliar symptom of oxygen deficiency is the yellowing of the main veins, followed by interveinal chlorosis of the young foliage and a major leaf drop of the more mature leaves.

Selected References

Seeley, J. G. 1949. The response of greenhouse roses to various oxygen concentrations in the substratum. Proc. Am. Soc. Hortic. Sci. 53:451-465.

White, J. W. 1976. Greenhouse roses: Diagnosis and remedy of nutritional disorders. Roses, Inc., Haslett, MI. 46 pp.

Air Pollution

Fluoride

Fourteen cultivars of garden roses exposed to controlled fumigations with fluoride concentrations of 1-3 ppb (parts per billion) for six months developed symptoms ranging from mild interveinal chlorosis to severe marginal necrosis of the leaves. The plants developed more than the usual number of branches, which were thinner than normal, so that plants were more compact than normal and somewhat dwarfed (Plate 69). Total dry weight was significantly decreased as a result of the treatment. The amount of fluoride uptake and the development of foliar injury symptoms varied widely among the different cultivars tested. No injury to the flowers was noted during the course of the experiment.

Selected Reference

Brewer, R. D., Sutherland, F. H., and Guillemet, F. B. 1967. The relative susceptibility of some popular varieties of roses to fluoride air pollution. Proc. Am. Soc. Hortic. Sci. 91:771-776.

Ethylene

Roses are classified as being "sensitive" to injury from ethylene gas. Typical symptoms include epinasty and curling of the youngest leaflets on the stem and yellowing along the principal veins of the older leaves, followed by abscission of leaves and leaflets (Plate 70). Stem elongation is slowed compared to that occurring on unexposed plants. Ethylene appears to cause some decrease in apical dominance and often results in the growth of more than the normal number of axillary buds from the branches of affected plants. Abortion of the terminal flower bud has been reported to occur following exposure of the plant to excessively high concentrations of ethylene (500 ppb).

Selected Reference

Piersol, J. R. 1974. Effect of ethylene on rose growth. Colo. Flower Growers Assoc. Bull. 286:5-6.

Mercury Vapor

During the mid-1950s, mercury toxicity caused a rash of losses of Better Times and Briarcliff roses in greenhouses in the northeastern United States. In each case, the greenhouse in which the plants had been growing had been painted earlier in the summer with a greenhouse paint formulated using a mercury-containing fungicide to protect the paint film from fungus attack after it was applied to the roof bars. The chemical used was di(phenylmercuric) dodecenyl succinate. As was later shown by controlled experimentation, the chemical decomposed under the conditions of use, releasing mercury vapor into the greenhouse atmosphere. When ventilation was reduced at the beginning of the heating season, the concentration of mercury vapor became great enough to cause severe toxicity to the roses. The release of mercury vapor could be prevented by overpainting with a mixture of five parts of dry lime sulfur, 10 parts of wheat flour, and 100 parts of water.

The symptoms of mercury injury to the susceptible cultivars included the production of off-color flowers; petals on these flowers ranged from bluish to pink or white or brown instead of the normal red or pink color. Young buds often failed to open, and the bud scales turned brown. In such cases, the entire bud could be easily lifted off the receptacle. Continued exposure of the plants to mercury vapor resulted in slow growth and reduced flower production.

Selected References

Dimond, A. E., and Stoddard, E. M. 1955. Toxicity to greenhouse roses from paints containing mercury fungicides. Conn. Agric. Exp. Stn., New Haven, Bull. 595. 19 pp.

Zimmerman, P. W., and Crocker, W. 1934. Plant injury caused by vapors of mercury and compounds of mercury. Contrib. Boyce Thompson Inst. 6:167-187.

Paint Volatiles

Volatile materials from certain paints may cause serious injury to roses. Paints containing xylene, naphtha, and mineral spirit cause necrosis and distortion of rose leaves, and lower leaves readily abscise. Flowers develop necrotic spots and premature bluing.

Selected Reference

Seeley, J. G. 1976. Some paints can cause plant injury. N.Y. State Flower Ind. Bull. 72:1-2, 5.

Pesticide Toxicity

Trifluralin Injury

Volatile herbicides in the atmosphere can cause foliar injury to roses. Trifluralin (Treflan and other trade names), used around greenhouse roses at application rates of 1-2 lb/A, caused the formation of very small leaves in those portions of stems where the leaves were very small and immature at the time of a three-day exposure (Plate 71). The injury resembled copper deficiency. Leaves above and below the affected area and the flower buds that developed later were unaffected by such short-term exposures. Symptoms developed in a commercial greenhouse about the time in the fall that ventilators were closed and artificial heating began, after Treflan had been applied earlier to control weeds in the growing benches.

Cyhexatin Injury

Cyhexatin (Plictran, Dowco 213), an effective miticide on many horticultural crops, appears to be highly phytotoxic to most commercial rose cultivars at application rates considerably below those normally considered necessary for effective mite control. California research workers reported severe leaf injury to the new growth (leaves just turning from red to green) on 14 commercial cultivars treated with 3 oz/100 gal of the 50% WP formulation of Plictran. Additional tests using 0.5 and 1.5 oz/100 gal also resulted in moderate injury to Forever Yours roses.

Selected References

Sciaroni, R. H., and Enomoto, R. A. 1974. Plictran on greenhouse roses. Florists' Rev. 153(3975):29.

Seeley, J. G., Bing, A., and Krapes, M. 1974. Treflan causes a small-leaf condition in greenhouse roses. Florists' Rev. 154(3990):36, 73.

Nutritional Deficiencies

Nitrogen Deficiency

Leaves of nitrogen-deficient plants show an overall yellow-green color. The older leaves are more severely affected than the young growth (Plate 73) and may turn completely yellow and abscise. Leaves are generally stunted, and internode length and stem diameter are reduced. Flowers of darker colored cultivars may appear several shades lighter than normal.

Phosphorus Deficiency

The first symptom of phosphorus deficiency is an overall stunting of leaves and shoot growth. Later, the older leaves may lose their luster, become gray-green, and drop off without turning yellow (Plate 74). Root development is reduced, which results in poor flower production and slow development of buds. A slight purpling of the underside of the midrib develops on some cultivars. This should not be confused with the healthy, reddish purple coloration of vigorously developing young shoots commonly observed in many cultivars. Flower petals of pink cultivars may become dark pink.

Potassium Deficiency

Reported descriptions of potassium deficiency vary considerably, depending on the cultivars studied and the growing media used in the study. Most authors, however, have found that potassium deficiency results in stunted growth; flower stems shorter than normal; small, short, deformed flower buds; and tip and marginal yellowing, browning, and necrosis of the older, lower leaves (Plate 75). Some authors have felt that potassium deficiency caused an increase in the incidence of blind shoots; others have found no direct relationship between the two problems.

Calcium Deficiency

At very low calcium concentrations in solution culture, rose roots become short, thick, and brittle and eventually die and turn black. Young leaves are distorted, and older leaves become dull gray-green and may bend down at their margins. Later, the edges of the leaves turn yellow and brown, and the discolored spots may coalesce into large necrotic blotches (Plate 76).

Specific studies have been conducted of the proper calcium-boron balance for roses. The basic symptoms described in the preceding paragraph occurred when both calcium and boron levels were low. As calcium levels were increased with medium levels of boron, most of the symptoms disappeared. As boron levels were increased with medium levels of calcium, abortion of the terminal buds ceased, but the distortion of the young leaves continued until calcium levels were also raised to the highest concentration used. It appears that both calcium and boron must be present in adequate quantities but also must be in the proper balance in relation to each other if normal plant growth is to occur.

Magnesium Deficiency

Magnesium deficiency shows up primarily on lower, older leaves. Chlorosis of the older leaves begins in the interveinal areas and progresses to necrosis, at first of small spots, but later of larger and larger areas. Later, these necrotic areas may turn into dark brown or purplish blotches essentially covering the entire leaf (Plate 77). Overall growth is stunted.

Sulfur Deficiency

Sulfur deficiency has not been reported for roses grown in soil-containing media but has been experienced in nutrient solution culture systems. It starts as a slight interveinal chlorosis of the young leaves and may progress to an overall light yellow-green coloration of the entire new growth.

Iron Deficiency

The initial symptom of iron deficiency is an interveinal chlorosis of the young leaves; the main veins remain green (Plate 78). If the deficiency continues, the newly formed leaves may remain very small and eventually become completely pale yellow or almost white.

Most cases of iron deficiency are caused by some interference in the availability or uptake of iron from the growing medium, rather than by an actual shortage of iron in the soil. Poor aeration, overwatering, root-knot nematodes, high soluble salt levels, excessively high or low soil temperatures, or high concentrations of manganese, zinc, or phosphorus may precipitate iron deficiency symptoms in the upper foliage. For permanent alleviation of the problem, these underlying causes must be corrected. Foliar sprays or soil applications of iron in an available form may cause temporary improvement, but the symptoms return rather quickly if the primary problem has not been removed.

Copper Deficiency

Copper deficiency is usually associated with the peat or muck type of growing media. The first symptoms include distorted young leaves with yellow tips that later become necrotic. The growing point dies, and short, stunted lateral shoots then develop. These symptoms may be confused with those caused by trifluralin injury.

Zinc Deficiency

Zinc deficiency is often precipitated by excessive liming of the growing medium. It may occur naturally in heavily leached, acid, sandy soils. Zinc deficiency symptoms are almost identical to those caused by copper deficiency, except that in zinc deficiency, the lateral shoots that develop after the meristem dies remain severely stunted, causing a symptom called "little leaf" or "rosette" (Plate 79). Both deficiencies may occur simultaneously and can be diagnosed conclusively only after leaf tissue analysis.

Boron Deficiency

White-flowered rose cultivars frequently show symptoms of boron deficiency before cultivars of other colors. When roses were grown completely in nutrient solution culture, boron deficiency symptoms showed up as necrosis of the growing point, followed by the development of lateral shoots and death of the meristem, followed by further development of more lateral shoots—i.e., the "witches'-broom" type of symptom.

For plants grown in soil, different kinds of symptoms have been reported. Petal margins of white and yellow cultivars became distorted and brown, and at times, flowers became completely necrotic. At other times, incidence of bullheads (i.e., flowers with markedly shortened petals that are abnormally thick and have margins that roll in; see Bullheads) was greater. Because a specific balance between boron and calcium is necessary for normal growth, if calcium levels are high, boron levels must also be high, or boron deficiency symptoms are likely.

Molybdenum Deficiency

Symptoms of molybdenum deficiency resemble those of moisture stress, namely, browning and necrosis of the tips and edges of leaves. Violet flecks sometimes appear on the parts of the leaves that are still living.

Selected References

Asen, S., and Tukey, H. B. 1953. Leaf scorch on the Snow White variety of greenhouse rose as influenced by various concentrations of boron and calcium. Proc. Am. Soc. Hortic. Sci. 61:515-522.

Carlson, W. H., and Bergman, E. L. 1966. Tissue analysis of greenhouse roses (*Rosa hybrida*) and correlation with flower yield. Proc. Am. Soc. Hortic. Sci. 88:671-677.

Oertli, J. J. 1966. Sulfur deficiency in rose plants. Florists' Rev. 138(3583):15.

Sadasivaiah, S. P., and Holley, W. D. 1973. Ion balance in nutrition of greenhouse roses. Roses Inc. Bull. Suppl., November. 27 pp.

Seeley, J. G., and Davidson, O. W. 1939. Interrelationships of calcium and phosphorus concentrations on the growth of roses. Proc. Am. Soc. Hortic. Sci. 37:967-972.

White, J. W. 1976. Greenhouse roses: Diagnosis and remedy of nutritional disorders. Roses Inc., Haslett, MI. 46 pp.

Nutritional Toxicities

Nitrogen Toxicity

Stunted growth and dark green foliage may result from excessive levels of nitrogen. If the soil is cold and poorly drained, excessive levels of ammoniacal or nitrite forms of nitrogen, both of which are toxic at relatively low concentrations, may accumulate. When soil moisture levels are very low, high nitrate levels may cause marginal necrosis of the leaves, similar to symptoms caused by high soluble salt levels (osmotic stress).

Phosphorus Toxicity

Toxicity symptoms caused directly by high phosphorus levels are very rare, but excessive applications of fertilizers containing phosphorus may interfere with the availability of copper, iron, zinc, and calcium.

Potassium Toxicity and Sulfur Toxicity

Many fertilizers that contain potassium have a high "salt index"; that is, they contribute strongly to the osmotic stresses brought on by high levels of soluble salts.

Therefore, potassium toxicity symptoms are the same as those caused by high total soluble salt levels, namely, chlorosis and marginal leaf necrosis, root loss, and wilting of soft, succulent shoots.

Excessive applications of sulfate-containing fertilizers can also cause symptoms typical of high levels of total soluble salts.

Calcium Toxicity

Excessive calcium levels affect roses primarily by their adverse effects on the availability and uptake of other essential mineral elements. Overliming, for example, can be an important cause of iron deficiency chlorosis.

Magnesium Toxicity

High levels of magnesium are toxic to roses only when potassium and calcium levels are low (see Salinity).

Manganese Toxicity

The older mature leaves of Better Times roses affected by manganese toxicity show small black spots in the interveinal areas (Plate 81). Interveinal chlorosis of the young leaves, typical of iron deficiency symptoms, can also occur because a balance between iron and manganese is needed for normal growth (Plate 80).

Iron Toxicity

Iron levels may sometimes be excessive in nutrient solution culture and may cause symptoms of copper, manganese, or zinc deficiency because of the interrelationships between these nutrients and iron.

Copper Toxicity

Copper toxicity symptoms have not been described in recent literature. In the past, however, frequent and repeated use of copper-containing fungicides such as Bordeaux mixture often resulted in leaf abscission followed by excessive leaf drop.

Boron Toxicity

Levels of boron only slightly above normal can rather quickly cause marginal browning and necrosis of the older leaves (Plate 82). Abscission of leaflets follows, with the midrib often remaining attached to the plant. Boron toxicity symptoms may be confused with magnesium or calcium deficiency symptoms.

Zinc Toxicity

The first symptom of zinc toxicity is the appearance of light green, transparent, water-soaked areas along the veins of the leaflets. Later, the rest of the leaf turns yellow, then brown. Affected leaves abscise irregularly. Foliage drops only after the leaflets have turned completely brown.

Selected References

Asen, S., and Tukey, H. B. 1953. Leaf scorch on the Snow White variety of greenhouse rose as influenced by various concentrations of boron and calcium. Proc. Am. Soc. Hortic. Sci. 61:515-522.

Carlson, W. H., and Bergman, E. L. 1966. Tissue analysis of greenhouse roses (*Rosa hybrida*) and correlation with flower yield. Proc. Am. Soc. Hortic. Sci. 88:671-677.

Coorts, G. D. 1958. Excess manganese nutrition of ornamental plants. Mo. Agric. Exp. Stn. Res. Bull. 669. 35 pp.

Sadasivaiah, S. P., and Holley, W. D. 1973. Ion balance in nutrition of greenhouse roses. Roses Inc. Bull. Suppl., November. 27 pp.

Seeley, J. G., and Davidson, O. W. 1939. Interrelationships of calcium and phosphorus concentrations on the growth of roses. Proc. Am. Soc. Hortic. Sci. 37:967-972.

White, J. W. 1976. Greenhouse roses: Diagnosis and remedy of nutritional disorders. Roses Inc., Haslett, MI. 46 pp.

Salinity

Many rose growers must use high-salinity water for irrigation and in addition add specific soluble fertilizers to provide continuous liquid fertilization of their producing plants. If fertilizer salts are added to such solutions without consideration of the quality of the water being used, salinity problems may be encountered (Plate 83). Work at Colorado State University has shown that rose yield and quality can be reduced 10–50% any time that the total salt level in irrigation water exceeds 1,600 micromhos/cm, and that if the concentrations of the various nutrient ions are not physiologically balanced, yield reductions can begin at total salt readings of 1,300 micromhos/cm.

Among the anions that contribute to salinity problems, the bicarbonate ion is most commonly troublesome. It is frequently found at high levels in irrigation water supplies, and when present causes a lime-induced chlorosis of the foliage. Sulfates do not appear to have a toxic effect other than their contribution to the total salinity levels of the solutions. Chlorides are intermediate between bicarbonates and sulfates in their effects.

High magnesium levels, particularly when levels of potassium and calcium are not high enough to be physiologically balanced, result in high numbers of unsalable flowers because of extreme chlorosis of the foliage.

Selected References

Hughes, H. E., and Hanan, J. J. 1976. Effects of salinity in water supplies on rose production: Experiment 1. Colo. Flower Growers Assoc. Bull. 323. 4 pp.

Hughes, H. E., and Hanan, J. J. 1977. Effects of salinity in water supplies on rose production: Experiment 2. Colo. Flower Growers Assoc. Bull. 327. 3 pp.

Glossary

A—acre
C—centigrade or Celsius
cm—centimeter (0.39 in.; 1 cm = 10 mm)
g—gram (453.59 g = 1 lb)
gal—gallon
in.—inch (2.54 cm)
kg—kilogram (2.23 lb)
L—liter (1.06 quarts)
lb—pound
m—meter (39.37 in.)
mg—milligram
ml—milliliter (1 ml = 0.001 L)
mm—millimeter (10 mm = 1 cm)
µg—microgram (10^{-6} g)
µm—micrometer (10^{-6} m)
nm—nanometer (10^{-9} m)
oz—ounce
ppb—parts per billion
ppm—parts per million

abscise (n. abscission)—to fall off, as with leaves, flowers, fruits, or plant parts
abstriction—the separation and discharge of a part, as in the formation of spores or conidia in fungi
acaricide—chemical or physical agent that kills or inhibits the growth of mites
acervulus (pl. acervuli)—erumpent, saucer-shaped, cushion-like fruiting body of a fungus bearing conidiophores, conidia, and sometimes setae
achene—simple, dry, one-celled, one-seeded, indehiscent fruit
adventitious—produced in an unusual or irregular position, or at an unusual time of development
aeciospore—binucleate, dikaryotic spore produced in an aecium of a rust fungus
aecium (pl. aecia)—more or less cup-shaped fruiting structure in the rusts; a "cluster cup" that produces chains of aeciospores
aerosol—a suspension of colloidal particles in a gas
agar—a non-nitrogenous, gelatinous mixture of polysaccharides obtained from certain red algae and used extensively as a solidifying agent in laboratory culture media for bacteria, fungi, and tissue culture
anatomy—the branch of plant morphology that deals with the internal structure and form of plants
angiosperm—flowering plant; a plant bearing seeds that develop in an enclosed carpel
anion—negatively charged ion; contrasted with cation
anther—pollen-bearing portion of a flower
anthocyanin—natural blue, purple, or red pigments that are especially common in flower petals
anthracnose—a disease with limited necrotic lesions, caused by fungi that produce spores borne in acervuli
antibiotic—damaging to life; a chemical, usually of microbial origin, that inhibits or kills other microorganisms
antibody—a specific protein produced by an organism in response to an antigen
antigen—a substance that stimulates production of antibodies
antiserum—a serum containing antibodies
apex (adj. apical)—tip; part of root or shoot containing apical meristem
apical dominance—the influence exerted by a terminal bud in suppressing the growth of lateral buds
apothecium (pl. apothecia)—cuplike or saucerlike ascus-bearing fruiting body (ascocarp)
appressorium (pl. appressoria)—flattened swelling on a germ tube or hypha of parasitic fungi that adheres to the surface of the host before fungus haustoria penetrate host cell
ascocarp—sexual fruiting body (ascus-bearing organ) of an ascomycete; i.e., apothecium, perithecium, cleistothecium
ascomycete—fungus that produces sexual spores (ascospores) within an ascus
ascospore—spore borne in an ascus by "free cell formation"
ascus (pl. asci)—saclike or clavate cell containing ascospores (typically eight) and borne in an ascocarp
aseptate—without cross walls
asexual—vegetative; without sex organs, gametes, or sexual spores
atrophy—to progressively decline in size (of an organ or an entire body)
autocatalytic—pertaining to a chemical reaction that is speeded up by a self-produced substance
autoecious—pertaining to rust fungi that complete their life cycle on one host
auxin—a natural hormone that regulates plant growth particularly through cell elongation rather than cell division
avirulent—unable to cause disease; nonpathogenic
axenic—germfree; without the presence of another organism
axil—angle formed by the leaf petiole and the stem
axillary—pertaining to buds, branches, or meristems that occur in the axil of a leaf

bacilliform—shaped like a short, blunt, thick rod
bacterium (pl. bacteria)—minute, procaryotic organism that usually lacks chlorophyll and exists mostly as a parasite or saprophyte
bare-root—describing plants planted with no soil surrounding the roots
basal—located at or near the base of a structure
basidiomycete—fungus that forms sexual spores (basidiospores or sporidia) on a basidium
basidiospore—haploid spore produced on a basidium
basidium (pl. basidia)—short, club-shaped, haploid promycelium produced by basidiomycetes
bent neck—condition caused by loss in rigidity in pedicel below the opening flower bud
binucleate—having two nuclei
bioassay—a test involving the response of a living cell or organism to an artificial stimulus
biological control—disease or pest control through counterbalance of microorganisms and other natural components of the environment
biotype—subspecies group of organisms that differ in biochemical, physiologic, or behavioral properties; group of individuals with like genetic makeup
blight—general term for sudden, severe, and extensive wilting

and/or death of leaves, stems, flowers, or entire plants
blind shoot—flower bud that fails to form and/or develop
blotch—irregularly shaped, usually superficial spot or blot
Bordeaux mixture—a mixture of copper sulfate, lime, and water used as a fungicide to control downy mildew and other fungi
budding—vegetative (asexual) reproduction of plants by grafting a stem bud of a desired species or cultivar onto the rootstock of another plant, often a different species or cultivar
bullhead—a flower bud with a lower than normal ratio of length to diameter, giving a flat-topped appearance to the bud, an increase in size and weight, an increase in the number of short petals, and a proliferation of secondary florets bearing mainly carpels near the base of the flower

callus—tissue composed of large, thin-walled parenchyma cells that develop on or below a wounded surface, often resulting in a firm thickening or protuberance; undifferentiated tissue (term used in tissue culture)
calyx (pl. calyxes)—a collective term for sepals; outermost flower whorl
cane—slender, hollow, jointed, usually flexible stem of certain plants such as rose
canker—a lesion on a stem with sharply limited necrosis of the cortical tissue
canopy—cover or horizontal projection of the vegetation of a plant formed by its leaves, branches, etc.
carpel—the ovule-bearing structure of a flower in angiosperms; often regarded as a single, modified, seed-bearing leaf
cation—a positively charged ion; in contrast to anion
chlamydospore—thick-walled or double-walled, asexual resting spore (terminal or intercalary) formed from hyphal cells or by transformation of conidial cells
chlorophyll—green, light-sensitive pigments, found chiefly in chloroplasts of higher plants, that participate in photosynthesis
chloroplast—organelle containing chlorophyll
chlorosis (adj. chlorotic)—failure of chlorophyll development caused by nutritional disturbance or disease; fading of green plant color to light green, yellow, or white
clavate—club-shaped, narrowing in the direction of the base
cleistocarp—cleistothecium
cleistothecium (pl. cleistothecia)—closed, usually spherical ascocarp; typical of the powdery mildews
clone—one of a group of genetically identical individuals resulting from asexual (vegetative) multiplication; any plant propagated vegetatively and therefore considered a genetic duplicate of its parent
collenchyma (adj. collenchymatous)—flexible supporting tissue composed of elongated living cells with unevenly thickened primary walls
color break—*See* flower break
conidiophore—simple or branched fertile hypha on which conidia are produced
conidium (pl. conidia)—asexual spore borne at the tip or side of a specialized hypha, called a conidiophore
corolla—collectively, the petals of a flower
cortex (adj. cortical)—tissues between the epidermis and phloem in stems and roots
crown—compacted series of nodes from which shoots and roots arise
cultivar—cultivated variety; group of closely related plants of common origin within a species that differ from other cultivars in certain minor details such as form, color, flower, or fruit
culture—artificial growth and propagation of organisms on nutrient media or living plants
cuticle—outer sheath or membrane of a nematode or plant
cytokinins—class of hormones important in many growth responses of plants, including cell division, cell enlargement, bud and root formation, breaking seed dormancy, flower photoperiod, and parthenocarpy
cytoplasm—living contents of a cell, except the nucleus

decline—condition characterized by general reduced vigor, dwarfed leaves and shoots, chlorosis, wilting, leaf drop, and reduced flower quality
desiccate—to dry up
dichotomous—branching, frequently successively, into two more or less equal arms
dicotyledonous—having two seed leaves
dieback—progressive death of shoots, leaves, or roots beginning at the tips
dikaryon (adj. dikaryotic)—fungal cell having two sexually compatible nuclei
diploid—having a double set of chromosomes per cell ($2n$)
disbud—to thin out flower buds to improve the quality of the bloom
disinfest—kill or inactivate disease organisms on surface of seed or plant part or in soil before they can cause infection
dissemination—spread of infectious material (inoculum) from a diseased to a healthy plant by wind, water, people, animals, insects, mites, machinery, etc.
diurnal—daily; during the day, as opposed to nocturnal
DNA—deoxyribonucleic acid
dodder—parasitic seed plant without leaves; a yellow, filamentous vine
dormant—resting; living in a state of reduced physiologic activity

electrophoresis—the differential movement of charged molecules in solution through a porous medium in an electric field; the porous supporting medium may be filter paper, cellulose acetate, or a gel
endemic—native to one country or geographic region
endoconidiophore—special hypha in which endoconidia are borne
endoconidium (pl. endoconidia)—conidium formed inside a hypha
endoparasite—parasite living inside its host
enzyme—protein that catalyzes a specific biochemical reaction
enzyme-linked immunosorbent assay (ELISA)—serologic procedure often used for virus assay
epidemic—general and serious outbreak of disease; used loosely of plants
epidemiology—study of factors influencing initiation, development, and spread of infectious disease
epidermis—surface layer of cells of leaves and other soft plant parts
epinasty—abnormal twisting and bending of stems and downward bending of leaves
epiphyte—plant living on another plant but not as a parasite
eradication—control of disease by eliminating the pathogen after it has become established
erumpent—bursting or erupting through the substrate surface
etiology—science or theory of the causes or origins of disease
extrude—to push out; emit to the outside

fascicle—small group, bundle, or cluster of flowers, leaves, stems, or roots; also used with fungi
fibrillose—covered with little fibers or elongated, generally thick-walled cells
filamentous or **filiform**—threadlike
fission—form of cell division; splitting into two by division of the complete organism
flagellum (pl. flagella)—hair-, whip-, or tinsel-like appendage of a motile cell (bacterium, zoospore) that provides locomotion
flexuous—having turns or bends; winding
flower break—break or stripe in flower color
foliar—pertaining to leaves
fruiting body—general term for complex, spore-bearing fungal structure
fumigate—to apply a vapor-active (volatile) disinfectant to kill microorganisms and other pests
fungicide (adj. fungicidal)—chemical or physical agent that kills or inhibits the growth of fungi
Fungi Imperfecti—group of fungi for which the sexual stage

is not known; also, the asexual stage of ascomycetes
fungus (pl. fungi)—organism lacking chlorophyll that reproduces by sexual or asexual spores and not by fission; generally speaking, a mycelium with well-marked nuclei
fusiform—spindlelike; narrowing toward the ends

gall—abnormal swelling or localized outgrowth, often more or less spherical, produced by a plant as the result of attack by a fungus, bacterium, insect, mite, or other agent
gametangium (pl. gametangia)—any cell or organ that produces gametes
gamete—sex cell; especially a cell formed in a gametangium that fuses with another sex cell in sexual reproduction
gene—smallest functional unit of genetic material on a chromosome; bearer of a hereditary trait
genome—set or group of chromosomes
genus (pl. genera)—group of related species
germinate—to begin growth (as of a seed, spore, sclerotium, or other reproductive body)
germ tube—hypha resulting from an outgrowth of the spore wall and/or cytoplasm
gibberellins—group of plant hormones important in many physiologic processes, including cell elongation, parthenocarpy, seed germination, and flowering
girdle—to remove a complete ring of bark (including phloem) from a tree or shrub; results in death from root starvation
globose—almost spherical
graft—transfer of aerial parts of one plant (e.g., buds or twigs) onto the root or trunk of a different plant
gymnosperm—vascular plant that produces seeds that are not enclosed within carpel tissues (naked seed)

haploid—having a single set of chromosomes
haustorium (pl. haustoria)—special hyphal branch of a parasite, especially one within a living host cell, for absorption of food; often associated with rusts, downy and powdery mildews, parasitic flowering plants, and other obligate parasites
helical—of, relating to, or having the form of a helix; spiral
herbaceous—describing plants that do not develop much woody tissue and thus remain soft and succulent
herbicide—chemical or physical agent that limits the growth of or kills plants
heteroecious—undergoing different parasitic stages on two unlike hosts, as in some rust fungi
heterothallic—having sexes separated in different mycelia
hip—rose fruit formed by a group of achenes surrounded by a receptacle and hypanthium
homology—likeness in structure
honeydew—sweet material exuded from leaves of many plants in hot weather; a honeylike secretion produced by many insects
hormone—organic chemical normally produced in minute amounts in one part of an organism and transported to another area of the same organism, where it affects growth and/or other functions
host—living plant attacked by (or harboring) a living parasite and from which the invader obtains part or all of its nourishment
host range—kinds of plants attacked by a given pathogen
hyaline—transparent or nearly so; frequently used in the sense of colorless
hybrid—offspring of two individuals of different genetic character
hypanthium—floral tube formed by the fusion of the basal portions of the sepals
hyperplasia (adj. hyperplastic)—excessive, abnormal, usually pathological multiplication of the cells of a tissue or organ
hypersensitive—displaying increased sensitivity, as when host tissue dies at the point of attack by a pathogen, so that infection does not spread
hypertrophy—excessive, abnormal, usually pathological enlargement of cells in a tissue or organ
hypha (pl. hyphae)—tubular filament of a fungal thallus or mycelium; the structural unit of fungi
hypoplasia (adj. hypoplastic)—excessive, abnormal, usually pathological multiplication of the cells of a tissue or organ

immune—not affected by or responsive to disease; exempt from disease
imperfect state—asexual portion of a life cycle of a fungus, during which asexual spores (such as conidia) or no spores are produced
indicator plant—plant that reacts to certain viruses, other pathogens, or environmental factors with specific symptoms; such plants are used to identify pathogens or determine the effects of environmental factors
infect (n. infection)—to enter and establish a parasitic relationship with a host plant
infection court—site in or on a host plant where an infection can occur
infectious—capable of infecting and spreading from plant to plant
infestation—attack by animals, especially insects or nematodes; aggregation of inoculum or other organisms on a plant surface
injury—result of transitory operation of an adverse factor such as insect feeding, action of a chemical, or adverse environmental factor
inoculate (n. inoculation)—to place inoculum (microorganism or virus, etc.) in an infection court
inoculum—pathogen or its parts, responsible for producing disease
insecticide—chemical or physical agent used to control or kill insects
intercalary cell—cell between two others
intercellular—between or among cells
internode—area between two adjacent nodes on a stem
interveinal—area between leaf veins
intracellular—within or through a cell or cells
in vitro—in glass or an artificial environment
in vivo—within a living organism
isolate—separated or confined spore or microbial culture; a fungus, bacterium, or other organism in pure culture
isometric—pertaining to a structural form that has three equal axes at right angles to one another

latent—present but not manifested or visible
leach—to wash soluble nutrients down through the soil
lenticel—pore in the bark of woody stems and other plant parts through which gases are exchanged
lesion—well-marked but localized diseased area
lignified—describing the impregnation of cellulose with lignin
lignin—complex organic substance that imparts rigidity and strength, especially to woody tissues

macroconidium (pl. macroconidia)—the larger, generally more diagnostic conidium of a fungus that also has microconidia
mechanical transmission (or inoculation)—spread or introduction of inoculum to an infection court by hand manipulation accompanied by physical disruption of the host tissue
meristem (adj. meristematic)—plant tissue that functions principally in cell division and differentiation; a mass of growing cells, capable of frequent cell division
mesophyll—chlorophyllous tissues of a leaf between the epidermal layers
metabolism—chemical transformation of nutrients into energy and by-products within a cell
metabolite—a product of metabolism
microconidium (pl. microconidia)—the smaller conidium of a fungus that also has macroconidia
microsclerotium (pl. microsclerotia)—microscopic, dense aggregate of darkly pigmented, thick-walled hyphal cells
midrib—central, thickened vein of leaves
mildew—plant disease characterized by a thin coating of mycelial growth and spores on the surfaces of infected plant parts
miticide—chemical or physical agent that kills or inhibits the growth of mites
MLO—mycoplasmalike organism
monocotyledonous—referring to a plant whose embryo has

one seedling leaf
mosaic—disease symptom characterized by a mottling of the foliage or by variegated patterns of dark and light green to yellow that form a mosaic; caused by disarrangement or unequal development of the chlorophyll content
mottle—disease symptom characterized by light and dark areas in an irregular pattern
muck—black soil containing decaying matter
mycelioid—resembling mycelium
mycelium (pl. mycelia)—mass of hyphae constituting the body (thallus) of a fungus
mycoplasma—procaryotic organism, smaller than conventional bacteria, lacking rigid cell walls and variable in shape

necrosis (adj. necrotic)—death of plant cells, usually resulting in tissue turning dark
nectary—gland at the base of a flower in which nectar is secreted
nematicide—chemical or physical agent that kills or inhibits nematodes
nematode—small, wormlike animal, parasitic in plants or animals or free-living in soil or water
node—region on the stem where leaves are attached; or the point of branching of the stem

obligate parasite—organism that can survive only on or in living tissue and that has not been cultured on laboratory media
oosphere—female gamete in some fungi
oospore—thick-walled resting spore that develops from an oosphere produced by fertilization or parthenogenesis
organelle—delimited, membranous structure within a cell having a specialized function
osmosis (adj. osmotic)—diffusion of a solvent (usually water) through a differentially permeable membrane, from the side of the higher concentration to the side of the lower concentration
ostiolate—having ostioles
ostiole—opening or pore, e.g., of a perithecium or pycnidium
overwinter—to survive the winter
ovoid—egg-shaped
ovule—enclosed structure that, after fertilization, becomes a seed

parasite (adj. parasitic)—organism living in or on another living organism (host) and obtaining food from it
parenchyma—physiologically active plant tissue composed of thin-walled cells that often store food and usually retain meristematic potentialities
parthenocarpy—natural or artificially induced development of fruit without sexual fertilization
parthenogenesis—development of an egg (female gamete) into a new individual without fertilization by a male sperm (male gamete)
pathogen—organism or agent that causes disease in another organism
pathogenicity—ability to cause disease
peat—partially decomposed plant tissue formed in water of marshes, bogs, or swamps, usually under conditions of high acidity
pedicel—small stalk; stalk of an individual flower
peduncle—stalk bearing a flower, flower cluster, or fruit
perfect stage (state)—stage in the life cycle of fungi in which sexual spores (e.g., ascospores and basidiospores) are formed after nuclear fission; sexual stage
perithecium (pl. perithecia)—flask-shaped or subglobose, thin-walled, fungus fruiting body (ascocarp) containing asci and ascospores; spores are expelled or otherwise released through a pore (ostiole) at the apex or tip
pest—any organism injuring plants or plant products
petal—one of the members of the corolla of a flower; frequently conspicuously colored
petaloid—resembling a petal in shape, texture, and/or color
petiole—the stem of a leaf; the stalk attaching a leaf blade to a stem
pH—measure of acidity—pH 7 is neutral; below pH 7 is acidic; above pH 7 is alkaline

phloem—food-conducting tissue in plants
photosynthesis—manufacture of carbohydrates from carbon dioxide and water in the presence of chlorophyll, using light energy and releasing oxygen
phyllody—change of flower parts to leaves
physiologic race—subdivision within a species; members are alike morphologically but differ from other races in virulence, symptom expression, biochemical and physiologic properties, or host range
phytopathology—plant pathology; science of plant disease
phytotoxic—harmful to plants
pistil—female reproductive organ of a flower
pitted—having depressions or cavities
plasmalemma—outer three-ply membrane bounding the protoplast next to the cell wall
plasmid—a small piece of DNA that carries small amounts of genetic information
pleomorphic—able to assume various shapes; having more than one independent form
ploidy—degree of repetition of the basic number of chromosomes in a cell, tissue, or organism
procaryotic (n. procaryote)—without internal membrane-bound organelles; lacking a nucleus
promycelium—basidium of the rusts that develops from the teliospore
propagation—reproduction by sexual or vegetative (asexual) means
propagule—any part of an organism capable of initiating independent growth
protective—referring to an agent, usually a chemical, that prevents or inhibits infection
protoplasm—living material of a cell
protoplast—plant cell exclusive of its wall
prune—to remove stems or branches of woody plants to control size and shape and improve quality and/or quantity of fruit and flowers
punky—describing decaying wood
pustule—small, blisterlike, frequently erumpent spot or spore mass
pycnidiospore—conidium produced within a pycnidium
pycnidium (pl. pycnidia)—asexual, flask-shaped, or globose fungus fruiting body containing conidia (pycnidiospores)
pycniospore—haploid, sexually derived spore formed in a pycnium
pycnium (pl. pycnia)—haploid, pycnidiumlike fruiting body or spermagonium produced by rust fungi
pyriform—pear-shaped

quarantine—legislative control of the transport of plants or plant parts to prevent spread of disease, pathogens, or other pests

race—subgroup or biotype within a species or variety, distinguished from other races by behavior (virulence, symptom expression, or host range) but not by morphology
receptacle—expanded upper end of pedicel to which flower parts are attached
recurved—curved backward or inward
resistance—ability of a host plant to overcome completely or to suppress, prevent, or impede the activity of a pathogen
respiration—series of chemical reactions whereby living protoplasm produces energy using oxygen, carbohydrate, and fat
ring spot—disease symptom characterized by yellowish or necrotic rings with green tissue inside the ring, as in some plant diseases caused by viruses
RNA—ribonucleic acid
rogue—to remove and destroy by hand individual plants that are diseased, infested by insects, or otherwise undesirable
rootstock—portion of the stem and associated root system onto which a scion is grafted
Rosaceae—plant family to which cultivated rose belongs
rosette—disease symptom characterized by short, bunchy growth habit caused by subnormal elongation of internodes
rugose—wrinkled; covered with coarse, netlike lines

salinity—the relative concentration of salts (especially sodium chloride) in water or soil
sanitation—destruction of infected and infested plants or plant parts
saprophyte (adj. saprophytic)—organism that feeds on dead organic matter
scion—portion of a stem that is transferred to a new rootstock in grafting
sclerotium (pl. sclerotia)—hard, frequently rounded, usually darkly pigmented resting body of a fungus, composed of a mass of specialized hyphal cells; the structure may remain dormant for long periods, then, when favorable conditions return, germinate to produce a stroma, fruiting body, mycelium, or conidiophores
scorch—"burning" of plant tissue from infection, lack of some element, chemical injury, or weather conditions
senescence—decline or degeneration, as with maturation, age, or disease stress
sepal—one of the outermost, sterile appendages of a flower that normally enclose the other floral parts in the bud; one of the separate parts of the calyx
septate—having cross walls
septum (pl. septa)—cross wall in a hypha
serology—study, detection, and identification of antigens, antibodies, and their reactions
sessile—having no stem
seta (pl. setae)—stiff, hairlike appendage; usually dark and thick-walled
shot hole—disease symptom where small, round fragments drop out of leaves, making them appear as if riddled by shot
solution culture—aqueous nutrient solution containing essential elements for cultivating plants without soil; hydroponics
sooty mold—dark fungus usually growing in insect honeydew or other high-carbohydrate substance
sorus (pl. sori)—compact fruiting structure
speciation—the process of differentiation into new species
spermagonium (pl. spermagonia)—flask-shaped, walled fungus structure in which spermatia are produced; also, a pycnium of a rust fungus
spermatium (pl. spermatia)—nonmotile, uninucleate haploid "sex" cell (haploid gamete), variously regarded as a pycniospore or microconidium
spiroplasma—procaryotic organism, smaller than conventional bacteria, lacking rigid cell walls; helical, flexuous mycoplasma
sporangiophore—differentiated hypha that bears a sporangium
sporangiospore—spore that develops in a sporangium
sporangium (pl. sporangia)—saclike or flasklike structure whose contents are converted into asexual spores (zoospores, sporangiospores)
sporulate—to produce spores
sterilization—method of destroying all microorganisms by heating to 100°C for 20 minutes
stipule—small structure or appendage found at the base of some leaf petioles
stock—portion of the stem and associated root system onto which a scion is grafted; also a term sometimes used for artificial breeding group and for a production planting
stoma or stomate (pl. stomata or stomates)—minute pore in the epidermis of leaves or stems that functions in gaseous exchange between a plant and its environment
strain—biotype; race; an organism or group of organisms that differs in minor aspects from other organisms of the same species or variety
streak—elongated lesion with irregular sides
stroma—compact mass of specialized hyphae in or on which fruiting bodies, spores, or both are produced
stunted—unthrifty; reduced in size and vigor because of unfavorable environmental conditions or a wide range of pathogens or abiotic agents
stylet—pointed, slender structure in mouth portion of plant-parasitic nematodes
subcuticular—beneath a cuticle
subelliptical—somewhat elliptical

subglobose—almost spherical
substrate—surface or medium on or in which an organism is living and from which it gets its nourishment
succulent—juicy; fleshy in texture or appearance
surfactant—monomolecular compound used as a detergent that reduces surface tensions and provides spreading action when used with pesticides
susceptible—not immune; lacking resistance; prone to infection
symptom—plant reaction that indicates disease
systemic—pertaining to chemicals or pathogens (or single infections) that spread generally throughout the plant body as opposed to remaining localized

tannin—one of a heterogeneous group of astringent phenol derivatives that are widely distributed in plants
teliospore—thick-walled resting spore produced by some fungi, notably rusts and smuts, that germinates to form a basidium
telium (pl. telia)—sorus producing teliospores
tendril—long, slender, coiling, modified leaf or structure associated with fungus spores
tetraploid—describing an organism that has twice the usual, diploid number of chromosomes
thallus—fungus body
thermal inactivation point—temperature at which virus particles are inactivated or lose their infectivity
tissue analysis—analysis of leaf tissues for major and minor elements
tissue culture—the technique of cultivating cells, tissues, or organs in a sterile, synthetic medium
tolerant—capable of sustaining disease without serious damage or yield loss
toxicity—capacity of a substance to produce injury
transmission—spread of virus or other pathogen from plant to plant
transpiration—loss of water vapor from aerial parts of plants, chiefly through stomata in the leaves
tumor—mass of tissue that grows independently of surrounding tissues and may invade those tissues
turgid—swollen or plump as a result of internal water pressure
turgor—inflation of a plant cell by the fluid contents

understock—portion of the stem and associated root system onto which a scion is grafted
unicellular—one-celled
uninucleate—having one nucleus
urediospore—binucleate, dikaryotic, asexual, one-celled repeating spore of rust fungi; borne in a uredium
uredium (pl. uredia)—fruiting body (sorus) of rust fungi that produces urediospores

vascular—pertaining to conductive (xylem and phloem) tissue
vector—agent (insect, mite, animal, human, etc.) able to transmit a pathogen (virus, bacterium, fungus, mycoplasma, nematode)
veinbanding—symptom of virus disease in which regions along veins are darker green than the tissue between veins
veinclearing—disappearance of green color in or along the leaf veins
vermiform—wormlike
vernalization—natural or artificial induction of early flowering by exposure of seeds to low temperatures; also, the requirement for breaking bud dormancy of certain temperate woody perennials
viability (adj. viable)—state of being alive; ability of seeds, fungus spores, sclerotia, etc., to germinate
virion—complete virus particle
viroid—smallest known agent of infectious disease, containing a small bit of RNA but no protein
virulence—degree or measure of pathogenicity; relative capacity to cause disease
virulent—highly pathogenic
virus—submicroscopic, filterable agent that causes disease and multiplies only in living cells and contains nucleic acid

surrounded by a protein coat

virus index—assay of plant tissues for presence of virus

volatile—evaporating or vaporizing readily

water-soaked—wet and dark and usually sunken and transparent

wilt—lack of freshness or drooping of leaves from lack of water (inadequate water supply or excessive transpiration); a vascular disease that interrupts the normal uptake and distribution of water

witches'-broom—disease symptom with an abnormal, massed, brushlike development of many weak shoots arising at or close to the same point

wound—an injury to a plant caused by cutting, scraping, or other external force

xylem—water-conducting tissue in plants

zoospore—asexually produced fungus spore having cilia or flagella and capable of locomotion in water

Index

Actinonema rosae, 8
Aeration, and iron deficiency, 37
Agrobacterium, 2
   radiobacter strain K84, 24
   rhizogenes, 25, 26
   tumefaciens, 23, 24, 25
Agrocin 84, 24
Air pollution, 35–36
   ethylene, 35
   fluoride, 35
   mercury vapor, 36
   paint volatiles, 36
Algal leaf and stem spot, 22
Alphitomorpha pannosa, 5
Alternaria spp., 21
   alternata, 21
   brassicae var. *microspora*, 21
Aphelenchoides, 31
Apple mosaic virus, 26, 27
Arabis mosaic virus, 26, 27
Armillaria mellea, 22
Asteroma rosae, 8
Austrian briers, 10

Baby ramblers, 1
Baccara roses
   bullheads in, 33
      control of, 34
   petal blackening in, 34
   petal bluing in, 34–35
Bacillus tumefaciens, 24
Bacteria
   diseases caused by, 23–26
      crown gall, 23–25
      hairy root, 25–26
   general description of, 2
Bacterium tumefaciens, 24
Bayse No. 3 (rootstock), 24
Begonia richmondensis, 26
Belonolaimus, 31
Bengal roses, 1, 28
Bent neck, 34
Beta vulgaris, 26
Bipolaris (Helminthosporium) setariae, 22
Black mold, 17–18
Black spot, 7–11
   disease cyle, 9
   resemblance to spot anthracnose, 20
Blind shoots. *See* Blindness
Blindness, 33
   relationship to potassium deficiency, 37
Blossom blight (*Botrytis*), 18
Blotch (black spot), 7
Bordeaux mixture, and copper toxicity, 38
Boron deficiency, 37
   relationship to bullheads, 37

Boron toxicity, 38
Boron-calcium balance, 37
Botryosphaeria ribis, 19
   var. *chromogena*, 19
Botrytis, 18, 19, 22
   cinerea, 18, 19
Botrytis blight, 18–19
Brand canker, 14, 15
Brandfleckenkrankheit, 15
Briarcliff (cultivar), 28
Bristle root, 25
Brooks 48 (rootstock), 24
Brown canker, 16–17
Bud and flower blight (*Botrytis*), 18
Bullheading. *See* Bullheads
Bullheads, 33–34
   relationship to boron deficiency, 37

Calcium deficiency, 37
Calcium toxicity, 38
Calcium-boron balance, 37
Cane blight (*Botrytis*), 18
Cane canker (*Botrytis*), 18
Cane canker (*Cylindrocladium*), 20
Canker, 19–20. *See also* Brand canker, Brown canker, Cane canker, and Common canker
Canker and gall (*Griphosphaeria*), 20
Catharanthus roseus, 27
Cephaleuros virescens, 22
Cercospora
   puderi, 21, 22
   rosicola, 21, 22
Cercospora leaf spot, 21
Chaenomeles japonica, 27
Chalaropsis thielavioides, 17
Chamaecyparis lawsoniana, 27
Champney roses, 1
Chenopodium
   amaranticolor, 27, 28
   murale, 28
   quinoa, 27, 28
China roses, 1, 28
Chrysanthemum
   frutescens, 26
   morifolium, 26
Clarke 1957 (rootstock), 24
Clitocybe tabescens, 22
Coleus blumei, 26
Colletotrichum capsici, 21
Columbia roses, bullheads in, 33
Common canker, 15–16
Common stem canker. *See* Common canker
Coniothyrium
   fuckelii, 15, 16
   rosarum, 15
   wernsdorffiae, 14, 15
Copper deficiency, 37
   resemblance to trifluralin injury, 36, 37

   resemblance to zinc deficiency, 37
Copper toxicity, 38
Corynebacterium, 2
Coryneopsis microsticta, 19
Cotoneaster parneyi, 26
Criconemoides, 31
Crown gall, 2, 20, 23–25
   biological control of, 24
   compared to hairy root, 23, 25
   hosts of, 23
   resemblance to galls caused by *Griphosphaeria corticola*, 20
Cryptosporella umbrina, 16, 17
Cryptosporium minimum, 19
Cucumis sativus, 27, 28
Curvularia brachyspora, 22
Cyamopsis tetragonoloba, 27
Cyhexatin injury, 36
Cylindrocladium scoparium, 20

Dagger nematode, 31, 32
Delphinium elatum, 26
Diaporthe
   eres, 20
   umbrina, 17
Didymella sepincoliformis, 20
Dieback, 19–20
Diplocarpon rosae, 8, 9
Disease control, general types of, 3. *See also* individual diseases
Ditylenchus, 31
Dr. Huey (rootstock), 12, 17, 29, 32
Dodder, role as a virus vector, 2
Dowco 213 injury, 36
Downy mildew, 13–14
   contrasted with powdery mildew, 13

Elaeagnus angustifolia, 26
ELISA
   technique, 26
   test, 27, 28
Elsinoë rosarum, 20
Environmental imbalances, 34–35
   heat and moisture stress, 35
   oxygen deficiency, 35
   petal blackening in Baccara roses, 34
   petal bluing in Baccara roses, 34–35
Eradication, disease control principle, 3
Erwinia, 2
Erysiphe, 5
   pannosa, 5
Ethylene injury, 35
Exclusion, disease control principle, 3

Floribunda, 1
Fluoride injury, 35

Fungi
  diseases caused by, 5–23
  general description of, 2

Gloeosporium
  rosaecola, 22
  rosarum, 22
Glomerella cingulata, 20, 22
Gray mold blight (*Botrytis*), 18
Gray mold rot (*Botrytis*), 18
Griphosphaeria corticola, 19, 20

Hairy root, 25–26
  compared to crown gall, 23, 25
Heat stress, 35
Helicotylenchus, 31, 32
  nannus, 31
Helminthosporium, 22
Hemicycliophora, 31
Hoplolaimus, 31
Hybrid perpetual roses, 1, 10, 28
Hybrid tea roses, 1, 10, 28, 29
Hybridization, 1

Impatiens hostii, 26
Infectious diseases, 2, 5–32
Iowa State University (ISU) 60-5
  (rootstock), 24
Iron deficiency, 37
  and overliming, 38
Iron toxicity, 38
IXL (rootstock), 10

Kalanchoë daigremontiana, 26

Leaf and stem nematodes, 31
Leaf blotch (black spot), 7
Leaf spot (black spot), 7. *See also*
  Cercospora leaf spot, Leaf spots
Leaf spots, 21. *See also* Cercospora leaf
  spot, Black spot
Leptosphaeria coniothyrium, 15
Lesion nematodes, 31, 32
Longidorus, 31
Lonicera japonica aureo-reticulata, 26
Lycopersicon esculentum, 26

Macroposthonia, 31, 32
  axeste, 31
Madame Butterfly (cultivar), 26, 27, 28, 29
Magnesium deficiency, 37
Magnesium toxicity, 38
  and salinity, 39
Malus sylvestris, 26, 27
Manetti, 1, 32. *See also Rosa manetti*
Manganese toxicity, 38
Marsonia rosae, 8
Marssonina rosae, 8, 10
Meloidogyne sp., 31, 32
  life cycle, 31
  hapla, 31, 32
Memorial rose, 1
Mercury vapor injury, 36
Mildew. *See* Downy mildew, Powdery
  mildew
Mineral spirit, in paint, 36
MLOs, 2

Moisture stress, 35
  resemblance to molybdenum deficiency, 37
Molybdenum deficiency, 37
  resemblance to moisture stress, 37
Momordica balsamina, 27
Monochaetia compta, 22
Moss roses, 10
Multiflora roses, 1, 28
Mycoplasmalike organisms, 2
Mycoplasmas, 2
Mycosphaerella
  rosicola, 21
  rosigena, 21

Naphtha, in paints, 36
Nectria cinnabarina, 20
Nematodes, 2, 30–32
  Aphelenchoides, 31
  Belonolaimus, 31
  Criconemoides, 31
  dagger, 31, 32
  diseases caused by, 30–32
  Ditylenchus, 31
  Helicotylenchus, 31, 32
  Hemicycliophora, 31
  Hoplolaimus, 31
  leaf and stem, 31
  lesion, 31, 32
  life cycle drawing, 31
  Longidorus, 31
  Macroposthonia, 31, 32
  Meloidogyne, 31, 32
  Neotylenchus, 31
  Paratylenchus, 31
  pin, 31
  Pratylenchus, 31, 32
  Psilenchus, 31
  ring, 31, 32
  root-knot, 30, 31, 32
  root-lesion, 30
  Rotylenchus, 31, 32
  sheath, 31
  spiral, 31, 32
  stem and bulb, 31
  sting, 31
  stubby root, 31
  stunt, 32
  stylet, 32
  Trichodorus, 31
  Tylenchorhynchus, 31, 32
  Tylenchus, 31
  Xiphinema, 31, 32
Neotylenchus, 31
Nicotiana
  clevelandii, 27
  rustica, 28
  tabacum, 26, 27, 28
Nitrogen deficiency, 36
Nitrogen toxicity, 38
  resemblance to osmotic stress, 38
Noisette roses, 1, 28
Noninfectious diseases, 2–3, 33–39
  air pollution, 35–36
    ethylene, 35
    fluoride, 35
    mercury vapor, 36
    paint volatiles, 36
  causes of, 2–3
  environmental imbalances, 34–35
    heat and moisture stress, 35

oxygen deficiency, 35
petal blackening in Baccara roses, 34
petal bluing in Baccara roses, 34–35
nutritional deficiencies, 36–38
nutritional toxicities, 38
pesticide toxicity, 36
physiologic problems, 33–34
  bent neck, 34
  blindness, 33
  bullheads, 33–34
salinity, 39
Nutritional deficiencies, 36–38
  boron deficiency, 37
  calcium deficiency, 37
  copper deficiency, 37
  iron deficiency, 37
  magnesium deficiency, 37
  molybdenum deficiency, 37
  nitrogen deficiency, 36
  phosphorus deficiency, 36–37
  potassium deficiency, 37
  sulfur deficiency, 37
  zinc deficiency, 37
Nutritional toxicities, 38
  boron toxicity, 38
  calcium toxicity, 38
  copper toxicity, 38
  iron toxicity, 38
  magnesium toxicity, 38
  manganese toxicity, 38
  nitrogen toxicity, 38
  phosphorus toxicity, 38
  potassium toxicity, 38
  sulfur toxicity, 38
  zinc toxicity, 38

Ophelia (cultivar), 26, 27, 28
Osmotic stress. *See* Salinity
Overliming, and iron deficiency
  chlorosis, 38
Overwatering, and iron deficiency, 37
Oxygen deficiency, 35

Paint volatiles, 36
Paratylenchus, 31
Pelargonium hortorum, 26
Pellicularia koleroga, 22
Pernetiana roses, 10
Peronospora sparsa, 13
Pesticide toxicity, 36
  cyhexatin injury, 36
  trifluralin injury, 36
Petal blackening, in Baccara roses, 34
Petal bluing, in Baccara roses, 34–35
Petal spots, 22
Petunia hybrida, 27, 28
Pezizella oenotherae, 22
Phaseolus vulgaris, 26, 27
Phomopsis mali, 20
Phosphorus deficiency, 36–37
Phosphorus toxicity, 38
Phragmidium, 11
  americanum, 11
  fusiforme, 11
  montivagum, 11
  mucronatum, 11
  rosae-californicae, 11
  rosae-pimpinellifoliae, 11

rosicola, 11
speciosum, 11
tuberculatum, 11
Phyllosticta rosae, 22
Phymatotrichum omnivorum, 22
Physalospora
   fusca, 22
   obtusa, 22
Physiologic problems, 33–34
   bent neck, 34
   blindness, 33
   bullheads, 33–34
Phytomonas
   rhizogenes, 25
   tumefaciens, 24
Phytophthora megasperma, 21
Pin nematodes, 31
Plictran injury, 36
Polyantha roses, 1, 10, 28
Potassium deficiency, 37
   relationship to blind shoots, 37
Potassium toxicity, 38
Powdery mildew, 5–7
   contrasted with downy mildew, 13
   disease cycle, 6
Pratylenchus, 31, 32
   life cycle, 31
   penetrans, 31, 32
   vulnus, 31, 32
Propagation of roses, general description of, 2
Protection, disease control principle, 3
Pruning, general description of, 2
Prunus
   mahaleb, 27
   persica, 27
   serrulata, 27
Prunus necrotic ringspot virus, 26, 27
Pseudomonas, 2
   tumefaciens, 24
Psilenchus, 31
Pyrus communis, 26

Queen Elizabeth (cultivar), 29

Ragged Robin (rootstock), 10, 12, 17
Ramblers, 1
Ramularia macrospora, 22
Rapture (cultivar), 26, 27, 28
Resistance, disease control principle, 3
Ring nematodes, 31, 32
Root rots, 22
Root-knot nematodes, 30, 31, 32
   and iron deficiency, 37
Root-lesion nematodes, 30
Rootstocks, and nematode infestation, 30, 32
Rosa spp., 1, 13, 26, 32
   borboniana, 1
   californica, 13
   canina, 13, 30
   caudata, 10
   centifolia, 13
   chinensis, 1
   dilecta, 1
   engelmanii, 11
   foetida, 11
   indica, 13
      cv. Major, 32
   manetti, 1, 10, 12, 17, 24, 26, 29. *See also* Manetti
   multiflora, 1, 10, 12, 17, 20, 21, 24, 28, 29, 32
      cv. Burr, 28, 29
      cv. 60-5, 32
   noisettiana, 1
      cv. Manetti, 32
   odorata, 1, 10, 12, 17, 29, 32
      cv. Sweet, 29
   × *rehderiana*, 1
   rubiginosa, 13
   rugosa, 28
   setigera, 30
   spinosissima, 11
   suffulta, 11
   wichuraiana, 1
Rosaceae, 23
Rose
   Austrian briers, 10
   baby ramblers, 1
   Baccara. *See* Baccara roses
   Bengal, 1, 28
   Champney, 1
   China, 1, 28
   diseases, general discussion of, 2–3
   floribunda, 1
   groups, 1
   hybrid perpetual, 1, 10, 28
   hybrid tea, 1, 10, 28, 29
   hybridization, 1
   moss, 10
   multiflora, 1, 28. *See also Rosa multiflora*
   Noisette, 1, 28
   Pernetiana, 10
   polyantha, 1, 10, 28
   ramblers, 1
   rugosa, 10, 28
   tea, 1, 10, 28
   value of cut flowers, 1
   wichuraiana, 1, 10, 28
Rose Actinonema (black spot), 7
Rose flower break, 30
Rose flower proliferation, 30
Rose graft canker, 15
Rose leaf Asteroma (black spot), 7
Rose leaf curl, 29–30
   resemblance to rose wilt, 29
Rose mosaic, 26–27
Rose mosaic virus, 26, 27, 28, 30
Rose neck droop, 34
Rose ring pattern, 28–29
   resemblance to symptoms caused by rose mosaic virus, 28
Rose rosette, 28
Rose spring dwarf, 29
Rose streak, 28
Rose streak virus, 28
Rose wilt, 29
   resemblance to rose leaf curl, 29
   resemblance to Verticillium wilt, 12
Rotylenchus, 31, 32
Rubus ursinus var. *loganobaccus*, 26
Rugosa roses, 10, 28
Rust, 11–12

Salinity, 39
   and high magnesium levels, 39
   interrelationships with heat and moisture stress, 35
   and iron deficiency, 37
   resemblance to nitrogen toxicity, 38
Sclerotium rolfsii, 22
Sedum spectabile, 26
Septoria rosae, 22
Sheath nematodes, 31
Sphaceloma, 22
   rosarum, 20–21
Sphaerotheca, 5
   humuli, 5
   macularis, 5
   pannosa, 5, 6, 7
      var. *persicae*, 5
      var. *rosae*, 5
Spider mites, 7
Spiraea vanhouttei, 26
Spiral nematodes, 31, 32
Spiroplasmas, 2
Spot anthracnose, 20–21
Star sooty mold (black spot), 7
Stem and bulb nematodes, 31
Stem spot, 22
Sting nematodes, 31
Strawberry latent ringspot virus, 27–28
Stubby root nematodes, 31
Stunt nematodes, 32
Stylet nematodes, 32
Sulfur deficiency, 37
Sulfur toxicity, 38

Talisman roses, bullheads in, 33
Tea roses, 1, 10, 28
Texas Wax (rootstock), 10
Therapy, disease control principle, 3
Tobacco streak virus, 30
Torenia fournieri, 27
Treflan injury, 36
Trichodorus, 31
Trifluralin injury, 36
Tylenchorhynchus, 31, 32
Tylenchus, 31

Verticillium, 12, 30
   albo-atrum, 12, 29
   dahliae, 12
Verticillium wilt, 12–13
   resemblance to rose wilt, 12
Vicia faba, 26
Vigna sinensis, 27
Viroids, 2
Viruses
   diseases caused by, 26–30
      rose flower break, 30
      rose flower proliferation, 30
      rose leaf curl, 29–30
      rose mosaic, 26–27
      rose ring pattern, 28–29
      rose rosette, 28
      rose spring dwarf, 29
      rose streak, 28
      rose wilt, 29
      strawberry latent ringspot virus, 27–28
      tobacco streak virus, 30
   general description of, 2
   identification of, 2, 26
   transmission of, 2

Welch (rootstock), 10, 24
Wichuraiana roses, 1, 10, 28
Wilt, 21. *See also* Rose wilt and Verticillium wilt
Witches'-broom, in boron deficiency, 37
World Federation of Rose Societies, 1

World Rose Convention, 1

Xanthomonas, 2
Xiphinema, 31, 32
   life cycle, 31
   coxi, 27

   diversicaudatum, 27, 28, 31, 32
Xylene, in paints, 36

Zinc deficiency, 37
   resemblance to copper deficiency, 37
Zinc toxicity, 38